Microstructure and Function of Cells

Andreas Bubel

Illustrated by Cecilia Fitzsimons

ACKNOWLEDGEMENTS

We are indebted to many colleagues (as acknowledged in the legends) for providing, and giving permission to use, their electron micrographs. Special gratitude is extended to Drs G. Charles, T. Jenkins, R. L. Fletcher and M. Carter of Portsmouth Polytechnic for their practical help and encouragement during the genesis of this work and to Professor Gareth Jones for providing photographic and reprographic facilities.

In addition, appreciation is extended to both spouses for their never-failing support during the preparation of this book.

Figs 108, 109, 137, are reproduced by permission of Oxford University Press.

Figs 34, 103a, 105, 163, 164, are reproduced by permission of The Rockefeller University Press.

Figs 11, 13, 19, 33, 58, 70d, 75a, 79, 82a, 86, 87, 89, 113, 118, 119a, 127, 145, 148, 153, 176, 179, are reproduced by permission of Springer Verlag, Heidelberg.

Figs 3, 6, 7, 9, 21, 37, 39, 53, 54b, 64, 70, 71, 72, 95, 96, 97, 106, 114, 136, 139, 171, 172, are reproduced by permission of Longman Group UK.

Figs 12, 13, 24a 27, 28, 29, 31a,b, 32a,32c, 36, 40, 44c,d,e, 49, 51, 52, 55, 56, 62, 63, 65b–d, 68, 69, 73b,c, 74e, 75b,76, 77, 86a, 90, 103b, 104, 110, 111, 117,119b, 120, 122, 123, 124, 129, 130, 132, 134,138, 140, 158, 159, 160, 165, 168, 169, 170, 177, 178.

Fig 46, reproduced by permission of Biology of the Cell.

Figs 154, 155, 156 reproduced by permission of The Canadian Journal of Zoology.

MICROSTRUCTURE AND FUNCTION OF CELLS
Electron micrographs of cell ultrastructure

ANDREAS BUBEL

illustrated by
CECILIA FITZSIMONS

both of Portsmouth Polytechnic

ELLIS HORWOOD LIMITED
Publishers · Chichester

Halsted Press: a division of
JOHN WILEY & SONS
New York · Chichester · Brisbane · Toronto

First published in 1989
ELLIS HORWOOD LIMITED
Market Cross House, Cooper Street,
Chichester, West Sussex, PO19 1EB, England
The publisher's colophon is reproduced from James
Gillison's drawing of the ancient Market Cross, Chichester.

Distributors:

Australia and New Zealand:
JACARANDA WILEY LIMITED
GPO Box 859, Brisbane, Queensland 4001,
Australia

Canada:
JOHN WILEY & SONS CANADA LIMITED
22 Worcester Road, Rexdale, Ontario, Canada

Europe and Africa:
JOHN WILEY & SONS LIMITED
Baffins Lane, Chichester, West Sussex, England

North and South America and the rest of the world:
Halsted Press: a division of
JOHN WILEY & SONS
605 Third Avenue, New York, NY 10158, USA

South-East Asia
JOHN WILEY & SONS (SEA) PTE LIMITED
37 Jalan Pemimpin # 05–04
Block B, Union Industrial Building, Singapore 2057

Indian Subcontinent
WILEY EASTERN LIMITED
4835/24 Ansari Road
Daryaganj, New Delhi 110002, India

© **1989 A. Bubel and C. Fitzsimmons/**
     Ellis Horwood Limited
British Library Cataloguing in Publication Data
Microstructure and function of cells
1. Man. Cells & tissues. Microstructure
I. Bubel, Andreas   II. Fitzsimons, Cecilia
611'.018

Library of Congress Card No. 88–13177

ISBN 0–7458–0111–0 (Ellis Horwood Limited)
ISBN 0–470–21176–8 (Halsted Press)

Printed in Great Britain by Hartnolls, Bodmin

PREFACE

The study of cells in recent years has progressed rapidly and has become fundamental to the study of the structure and function of living organisms. In general, this interest has resulted in the publication of a large number of cell biology texts. These have mainly dealt with the cell organelles, tissues and organs of mammals, in terms of structure and function. This book, however, has been written to emphasize the wide structural variations that occur in cell organelles, cells and tissues of invertebrates and plants, in addition to those of mammals. The electron micrographs in this book have been selected to reflect this emphasis. Unlike other books, a collection of line drawings has been prepared especially to illustrate many electron micrographs. These have been included to help with interpretation of the electron micrographs. The content of the book is organized in a manner that is most useful to the student proceeding from simple to more complex matters. The text is an up-to-date brief account of the structure and function of organelles, cells and tissues in both animal and plants. Emphasis has been placed on present knowledge rather than on introductory historical background. It is hoped that the book will be useful to a variety of students in their academic pursuits and that it will serve to make their endeavours more enjoyable, complete and meaningful. It is intended primarily as a textbook for college and university courses in cell biology and for students who wish to gain a general view of modern cell biology for purposes of teaching or investigations in other fields, such as medicine, physiology, veterinary medicine, general biology, zoology and botany.

SOURCES OF ILLUSTRATIONS

U.S.A.

Dr K. M. Baldwin
College of Medicine, Dept Anatomy, Howard University, Washington D.C. 20059, U.SA.

Dr J. B. Delcarpio
Dept Anatomy, Louisiana State University Medical Center, 1901 Perdido Street, New Orleans, Louisiana 70112-1393, U.S.A.

Dr J. Cartwright Jr
Dept Pathology, Baylor College of Medicine, One Baylor Plasa, Houston, Texas 77030, U.S.A.

Prof. T. W. Keenan
Dept Biochemistry and Nutrition, Virginia Polytechnic Institute and State University, Blacksburg, Virginia 24061, U.S.A.

Dr K. Selman
College of Medicine, Dept Anatomy and Cell Biology, University of Florida, Gainesville, Florida 32610, U.S.A.

Dr D. Spangenburg
Dept Pathology, Eastern Virginia Medical School, Post Office Box 1980, Norfolk, Virginia 23501, U.S.A.

Dr K. J. Karnaky Jr
Dept Anatomy and Cell Biology, Medical University of South Carolina, 171 Ashley Avenue, Charleston, South Carolina 29425, U.S.A.

Dr E. A. Perkett,
Dept Medicine, School of Medicine, Vanderbilt University, Nashville, Tennessee 37232, U.S.A.

Dr W. H. Fahrenbach
Dept Electron Microscopy, Oregon Regional Primate Research Center, 505 Northwest 185th Avenue, Beaverton, Oregon 97006-3499, U.S.A.

Dr K. J. Eckelbarger
Harbor Branch Oceanographic Institute Inc., 5600 Old Dixie Highway, Fort Pierce, Florida 33450, U.S.A.

Prof. K. S. Kim
Dept Plant Pathology, 217 Plant Sciences Building, University of Arkansas, Fayetteville, Arkansas 72701, U.S.A.

Dr W. G. Jerome
Dept Pathology, The Bowman Gray School of Medicine, Wake Forest University, 300 South Hawthorne Road, Winston-Salem, North Carolina 27103, U.S.A.

Dr D. P. Dylewski
Dept Biochem. & Nutrition, Virginia Polytechnic, Blacksburg, Virginia 24061, U.SA.

Dr J. M. Burke
Dept Ophthalmology, Milwaukee County Medical Complex, 8700 West Wisconsin Avenue, Milwaukee, Wisconsin 53226, U.S.A.

Dr J. Thorsch
Dept Biological Sciences, University of California, Santa Barbara, California 93106, U.S.A.

Dr M. S. Forbes
Dept Physiology, University of Virginia, Box 449, Medical Centre, Jordan Medical Education Building, Charlottesville, Virginia 22908, U.S.A.

Dr E. LeCluyse
Interx Research Corp., Merck, Sharp & Dohme Research Labs, 2201 W. 21st St., Lawrence, Kansas 66046, U.S.A.

Dr D. G. Emery
Dept Zoology, Science Building 11, Iowa State University, Ames, Iowa 50011, U.S.A.

Dr T. Ichimura
Division of Biology and Medicine, Brown University, Providence, Rhode Island 02912, U.S.A.

Dr L. W. Wilcox
Dept Botany, Ohio State University, 1735 Neil Ave., Columbus, Ohio 43210, U.S.A.

Dr J. S. Ryerse
Pathology Dept, St. Louis University School of Medicine, 1402 So. Grand Blvd., St. Louis, Missouri 63104, U.S.A.

Dr S. E. Stevens
College of Science, Dept Molecular and Cell Biology, The Pennsylvania State University, Paul M. Althouse Lab., University Park, Pennsylvania 16802, U.S.A.

Dr S. Tyler
Dept Zoology, Murray Hall, University of Maine, Orono, Maine 04469-0146, U.S.A.

Dr S. Brocco
Dept Zoology, University of Washington, Seattle, Washington 98195, U.S.A.

Dr M. M. Friedman
Georgeton University, Dept Microbiology, Division of Molecular Virology and Immunology, 5640 Fishers Lane, Rockville, Maryland 20852, U.S.A.

Dr G. Reed
Dept Biological Sciences, Dartmouth College, Hanover, New Hampshire 03755, U.S.A.

Prof. O. C. McKenna
Dept Biology, The City College of the City University of New York, New York 10031, U.S.A.

Dr G. Olson
Dept Cell Biology, School of Medicine, Vanderbilt University, Nashville, Tennessee 37232, U.S.A.

Dr B. F. King
Dept Human Anatomy, University of California, Davis, California 95616, U.S.A.

Prof. T. J. Bradley
Dept of Development and Cell Biology, University of California, Irvine, California 92717, U.S.A.

Dr M. U. Nylen
Director, Extramural Program National Institute of Dental Research, Dept Health and Human Services, National Institute of Heath, Building WW, Room 503, Bethesda, Maryland 20205, U.S.A.

Dr J. E. Bodammer
U.S. Dept Commerce, National Oceanic and Atmospheric Administration, National Marine Fisheries Service, Oxford, Maryland 21654.

Dr P. H. De Bruijn
Dept Anatomy, The University of Chicago, 1025 East 57th Street, Chicago, Illinois 60637, U.S.A.

Prof. E. Anderson
Dept Anatomy and Cell Biology, Laboratory of Human Reproduction and Reproductive Biology, Harvard Medical School, 45 Shattuck Street, Boston, Massachusetts 02115, U.S.A.

Dr K. Esau
Dept Biol. Sciences, University of California, Santa Barbara, California 93106, U.S.A.

SOUTH AMERICA

Dr E. M. Rodriguez
Institute de Histologie y Patologia, Universidad Austral de Chile, Valdivia, Chile.

CANADA

Dr V.G. Vethamany
Audio-Visual Division, Faculty of Medicine, Sir Charles Tupper Medical Building, Dalhousie University, Halifax, N.S. B3H 4H7, Canada.

Dr R. R. Shivers
Dept Zoology, University of Western Ontario, London, Ontario N6A 5B7, Canada.

Dr K. Wright
Dept Zoology, Ramsay Wright Zoological Laboratories, University of Toronto, 25 Harbord Street, Toronto M5S 1A1, Ontario, Canada.

Dr M. Neuwirth
Electron Microscopy Section, Albert Environmental Centre, Bag 4000, Vegreville, Alberta T0B 4L0, Canada.

Dr R. F. Carter
Dept Pathology, Ontario Veterinary College, Geuelph, Ontario N16 2W1, Canada.

AUSTRALIA

Dr R. Smith
TV Science Unit, Australian Broadcasting Corporation, G.P.O. Box 9994, Sydney 2001, New South Wales, Australia.

Dr B. G. M. Jamieson
Zoology Dept, University of Queensland, St. Lucia, Brisbane 4067, Queensland, Australia.

Dr M. A. Borowitzka
School of Environmental and Life Sciences, Murdoch University, Murdoch, Western Australia 6150.

AFRICA

Prof. J. N. Maina
Dept Veterinary Anatomy, University of Nairobi, P.O. Box 30197, Nairobi, Kenya, Africa.

Dr X. Mattei
Dept Biologié Animalé, Faculté des Sciences, Université de Dakar, Dakar, Senegal, Africa.

JAPAN

Prof. S. Kanamura
Dept Anatomy, Kansai Medical University, Fumizono-Cho 1, Moriguchi, Osaka, 570 Japan.

Dr M. Fukumoto
Nagoya City University, College of General Education, Mizuho-Ku, Nagoya, 467 Japan.

Dr S. Sugiyama
Dept Dermatology, Sapporo Medical College, Minami 1, Nishi 16, Sapporo, Japan.

Dr T. Komuro
Dept Anatomy, Ehime University School of Medicine, Shigenobu, Shigenobu, Ehime, 791-02 Japan.

Dr T. Hatae
Dept Anatomy, Kagawa Medical School, Kagawa, 761-07 Japan.

EUROPE

Prof. Dr R. Dierichs
Anatomisches Institut, Westalische Wilhelms-Universitat, Vesaliusweg 2–4, D-4400 Münster, Fed. Rep. Germany.

Prof. Dr N. Paweletz
Institute of Cell and Tumour Biology, German Cancer Research Centre, D-6900 Heidelberg 1, Fed. Rep. Germany.

PD Dr W. Reisser
FB Biologie, Lahnberge, D-3550 Marburg, Fed. Rep. Germany.

Dr P. Siefert
Zoologisches Institut der Universitat, 8 Munchen 2, Luisenstrasse 14, Fed. Rep. Germany.

Dr T. A. Keil
Max-Planck-Institut fur Verhaltens-physiologie, Abteilung Schneider, D-8131 Seewiesen, Fed. Rep. Germany.

Prof. Dr E. Florey
Fakultat fur Biologie, Universitat Konstanz, Postfach 5560, D-7750 Konstanz 1, Fed. Rep. Germany.

Dr R. A. Steinbrecht
Max-Planck-Institut fur Verhaltens-physiologie, Abteilung Schnieder, D-8131 Seewiesen, Fed. Rep. Germany.

Dr G. Alberti
Zoologisches Institut 1 (Morphologie/Okologie), Universitat Heidelberg, Im Neuenheimer Feld 230, D-6900 Heidelberg, Fed. Rep. Germany.

Prof. Dr V.Storch
Zoologisches Institut der Universitat Heidelberg, Morphologie/Okologie, 6900 Heidelberg 1, Im Neuenheimer Feld 230, Fed. Rep. Germany.

Prof. Dr J. Bereiter-Hahn
Arbeitsgruppe Kinematische Zellforschung, Fachbereich Biologie, Johann Wolfgang Goethe-Universitat, Postfach 11 19 32, Senckenberganlage 27, D-6000 Frankfurt am Main, Fed. Rep. Germany.

Dr H-J. Beckmann
Anatomisches Institut der Universitat Munster Versaliusweg 2-4, D-4400 Munster, West Germany.

Prof M. Derenzini
Dipartimento di Patologia Sperimentale, Universita Delgi di Bologna, 40126 Bologna, Via S. Giacomo 14, Italy.

Dr R. Dallai
Dipartimento di Biologia Evolutiva, Universita di Siena, Via Mattioli 4, 53100 Siena, Italy.

Dr F. Guisti
Universita di Siena, Via Mattioli 4, 53100 Siena, Italy.

Dr M. Nigro
Instituto di Zoologia e anatomia Comparata, Universita di Pisa, 56100 Pisa, Via A. Volta 4, Italy.

Dr G. B. Martinucci
Dipartimento di Biologia, Universita di Padua, 35131 Padova, Italy.

Dr M. G. Selmi
Dipartimento di Biologia Evolutiva, Universita di Siena, Via Mattioli 4, 53100 Siena, Italy.

Dr L. Vitellaro-Zuccarello
Dipartimento di Fishiolgia e Biochimica Generale Sezione di Istologia e Anatomia Umana, Universita delgi Studi di Milano, Via Celoria 26, 20133 Milano, Italy.

Prof. E. Pannese
Istituto di Istologia, Embriologia e Neuoci-tologia, Universita delgi Studi di Milano, Via Mangiagalli 14, 20133 Milano, Italy.

Dr P. Willenz
Faculte des Sciences C.P. 160, Laboratoire de Biologie Animale et Cellulaire, Avenue F.D. Roosevelt 50, Universite libre de Bruxelles, B-1050 Bruxelles, Belgium.

Prof. Dr T. Sminia
Faculteit der Geneeskunde, Vakgroep Celbiologie Histologisch Laboratorium Vrije Universiteit, 1007 MC Amsterdam, Postbus 7161, Netherlands.

Dr M.-C. Lebart-Pedebas
   Faculte des Sciences, Laboratoire de Cytologie, Université Pierre et Marie Curie, 7 Quai St. Bernard, 75230 Paris Cedex 05, France.
Dr L. Ovtracht
   Laboratoire de Biologie Cellulaire, Université Paris VI, 67 Rue Maurice-Gunsbourg, Ivry, France.
Dr D. Georges
   Lab. Zool. Biol., Université Scientifique et Medicale, Grenoble, France.
Dr T. Haug
   Institut de Zoologie, Université de Neuchatel, CH-2000 Neuchatel, Switzerland.
Dr W, Witalinski
   Dept Comparative Anatomy, Jagiellonia University, Karasia 6, 30-060 Krakow, Poland.
Dr F. J. Medina
   Centro de Investigaciones Biologicas, Velazquez 144, 28006 Madrid, Spain.
Prof. M. M. Magalhaes
   Institute of Histology and Embryology, Faculty of Medicine Centre for Experimental Morphology of the University (INIC), Porto, Portugal.
Dr M. Holley
   Dept Physiology, The Medical School, University of Bristol, University Walk, Bristol BS8 1TD, U.K.
Dr G. Gabella
   Dept Anatomy, University College of London, Gower Street, London WC1E 6BT, U.K.
Dr T. Coaker
   Dept Histopathology, The Royal Victoria Infirmary, Queen Victoria Road, Newcastle upon Tyne NE1 4LP, U.K.
Dr D. N. Furness
   Dept Communications and Neuroscience, University of Keele, Keele, Staffordshire ST5 5BG, U.K.
Dr K. S. Richards
   Dept Biological Sciences, University of Keele, Keele, Staffordshire ST5 5BG, U.K.
Dr P. Newell
   School of Biological Sciences, Queen Mary College, University of London, Mile End Road, London E1 4NS, U.K.
Dr J. A. Firth
   Dept Anatomy and Cell Biology, St Mary's Hospital Medical School, Norfolk Place, London W2 1PG, U.K.

Dr S. Hunt
   Dept Biological Sciences, University of Lancaster, Lancaster, Lancashire LA1 4YQ, U.K.
Dr A. F. Rowley
   Dept Zoology, University College of Swansea, Singleton Park, Swansea SA2 88P, U.K.
Prof. G. Owen
   Principal, The University College of Wales, Aberystwyth SY23 2AX, Dyfed, U.K.
Dr R. L. Fletcher
   Dept Biological Sciences, Portsmouth Polytechnic, King Henry I Street, Portsmouth, U.K.
Dr G. Charles
   Dept Biological Sciences, Portsmouth Polytechnic, King Henry I Street, Portsmouth, U.K.
Dr T. Jenkins
   Dept Biological Sciences, Portsmouth Polytechnic, King Henry I Street, Portsmouth, U.K.
Dr A. Larkman
   Dept Biological Sciences, Portsmouth Polytechnic, King Henry I Street, Portsmouth, U.K.

CONTENTS

CELL STRUCTURE

The typical cell with a nucleus and cytoplasm and other organelles is not the smallest mass of living protoplasm: simpler or more primitive units of life exist. Unlike higher or eukaryotic cells, which have a true nucleus, primitive or prokaryotic cells, which comprise bacteria and blue-green algae, lack a distinct nucleus and contain a nuclear substance that is mixed, or is in direct contact, with the rest of the protoplasm (Fig 178ab).

The eukaryotic cell consists of a nucleus and a mass of cytoplasm surrounded by a plasma membrane. In multicellular organisms, the cells vary in shape and structure and are differentiated according to their specific functions in various tissues and organs. However, general characteristics common to all cells do exist. The shape of cells depends mainly on functional adaptations and partly on the surface tension and viscosity of the protoplasm, the mechanical action exerted by adjoining cells, and the rigidity of the plasma membrane. The original spherical shape of a cell is usually modified by its contact with other cells. The cells of many plants and animal tissues have polyhedral shapes determined by reciprocal pressures (Fig 103b). The sizes of different cells are found to range within broad limits. Some plant and animal cells are visible to the naked eye, e.g. eggs of some vertebrates have a diameter of several centimetres. However, the great majority of cells are only a few micrometres in diameter and are only visible with the microscope. In general, the volume of a cell is fairly constant for a particular cell type and is independent of the size of an organism. Kidney cells, for example, are about the same size in a horse and a mouse. The difference in total mass of an organ, in fact, depends on the number and not the individual volume of the cells.

When the cells of higher plants and animals are examined under the electron microscope, their complex structural organization is revealed. They are characterized by a nucleus bounded by a true nuclear envelope or membrane, and cytoplasm, limited by a plasma membrane (Fig 1a). In a plant cell, the plasma membrane is covered and protected on the outside by a thick cell wall (made of cellulose) through which there run channels, the plasmodesmata, by which adjacent cells communicate by means of fine cell processes (Figs 1bc, 177). In contrast, in animal cells, parts of the plasma membrane are covered by a thin layer of material which is described as the extraneous coat of the plasma membrane, i.e. glycocalyx on the apical plasma membrane and basement membrane (basal lamina) on the basal plasma membrane (Figs 19a, 22a, 73b).

The cytoplasmic component of cells is characterized by a prodigious development of membranes. The plasma membrane can be highly specialized and modified. Among such specializations are microvilli, cilia, invaginations and cellular attachment devices (Fig 1a). From the surface of most cells project a varied number of cellular extensions, collectively referred to as microvilli. They are usually devoid of cell organelles, except for a varied number of straight microfilaments present at their cores (Figs 14–20). Microvilli increase the cell surface available for exchange processes and are thought to participate in absorptive processes. Their microfilaments probably aid the transport of absorbed materials, in addition to their more obvious cytoskeletal (supportive) function. In wandering cells, microvilli are very likely instrumental in cellular movements. Cilia are motile cell projections. The ciliary cytoplasm contains a core of regularly arranged microtubules, nine pairs arranged around two single ones in a central position (Figs 65–72). Invaginations of the plasma membrane also greatly increase the surface of a cell and serve as an interface for enzymatic processes and transport across the membrane. In addition, drop-like invaginations of the plasma membrane may form either small pinocytotic vesicles or larger phagocytotic vesicles. Pinocytotic vesicles participate in the uptake of extracellular material (endocytosis) as well as in segregation or discharge of intracellular material (exocytosis) (Fig 63). When pinocytotic and phagocytotic invaginations break off from the plasma membrane into the cytoplasm they are referred to as phagosomes. A variety of plasma membrane specializations also provide cellular attachments between cells, as characterized by an epithelium. Variations in the size and shape of, intercellular junctions result in their classification as maculae (focal junctions), i.e. macula adhaerens, or zonulae (bands), i.e. zonula adhaerens. A gap junction

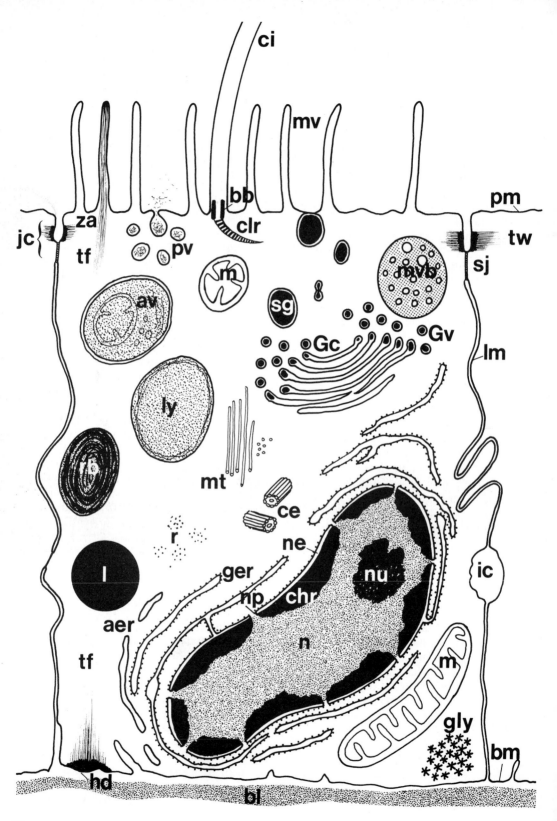

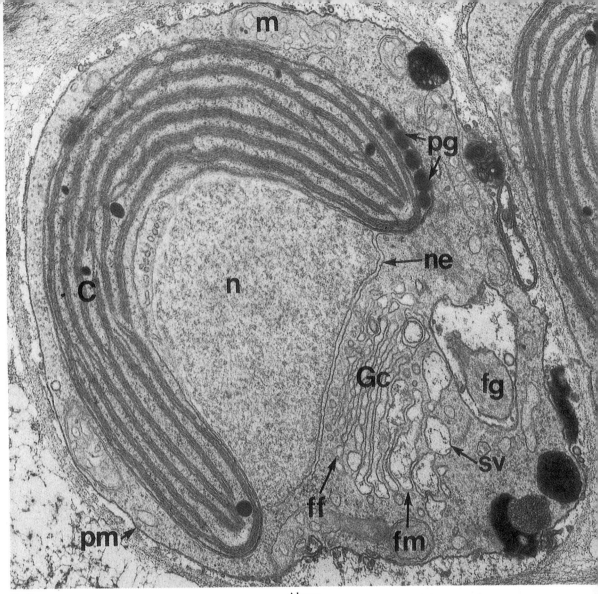

Left

Fig. 1(a) – General diagram of the ultrastructure of an animal cell.

aer, agranular endoplasmic reticulum; av, autophagic vacuole; bb, basal body; bl, basal lamina; bm, basal plasma membrane; ce, centriole; chr, chromatin; ci, cilium; clr, cilium rootlet; Gc, Golgi complex; ger, granular endoplasmic reticulum; gly, glycogen; Gv, Golgi vesicle; hd, hemidesmosome; ic, intercellular space (canal); jc, junctional complex; l, lipid; lm, lateral membrane; ly, lysosome; m, mitochondrion; mt, microtubules; mv, microvilli; mvb, multivesicular body; n, nucleus; ne, nuclear envelope; np, nuclear pore; nu, nucleolus; pm, plasma membrane; pv, pinocytotic vesicle; r, ribosome; rb, residual body; sg, secretory granule; sj, septate junction; tf, tonofilaments; tw, terminal web; za, zonula adhaerens.

Above

Fig. 1(b) – Electron micrograph of a spore of the alga, *Petelonia*, showing the main features of a plant cell. × 25 000. (Dr A. Bubel, Portsmouth Polytechnic.) C, chloroplast; Gc, Golgi complex; ff, forming face; fg, flagellum; m, mitochondrion, n, nucleus; ne, nuclear envelope; fm, maturing face; pg, pigment granule; sv, secretory vesicle.

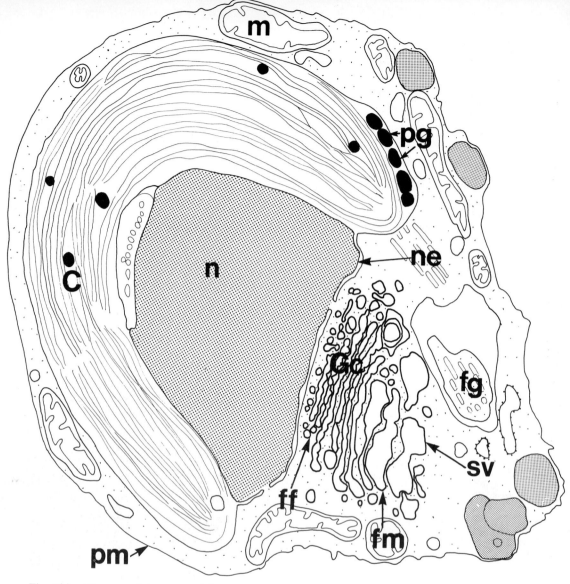

Fig. 1(c) – Diagram of the spore of the alga, *Petelonia.*
C, chloroplast; m, mitochondrion; fm, maturing face; pm, plasma membrane.

or nexus provides for ion communication between cells in many epithelia. A special type of structure, the hemidesmosome or half desmosome, is frequently found at the base of epithelial cells where they are anchored to the underlying connective tissue via a basal lamina, immediately adjacent to the basal plasma membrane (Figs 3–13). Microfilaments which originate in microvilli cores traverse the length of a cell and terminate in hemidesmosomes on the basal plasma membrane. These microfilament bundles are thought to act in a cytoskeletal (supportive) capacity (Figs 73–74). Under-

lying the basal membrane, the connective tissue is generally composed of collagen fibrils embedded in a ground substance (Fig 78).

A basic membranous organization is found in cell organelles such as mitochondria and lysosomes. Mitochondria are enclosed by two membranes, the inner one, which is folded, forms cristae that partially compartmentalize the mitochondrion (Figs 47–57). Mitochondria contain DNA, ribosomes and enzymes necessary for synthesizing a variety of proteins, many of which are involved in aerobic respiration. Lysosomes are membrane bound and

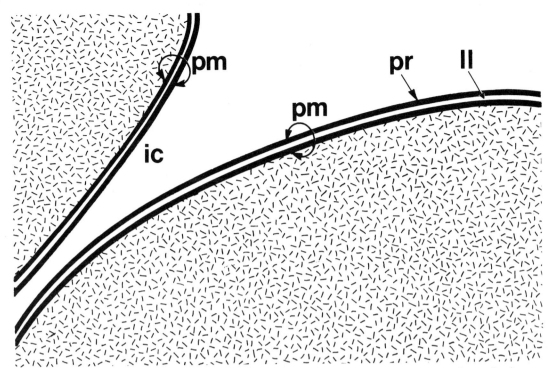

Fig. 2 – Diagram showing the three-layered structure (unit membrane) of a plasma membrane (pm). ic, intercellular space; ll, lipid layers; pr, protein layers.

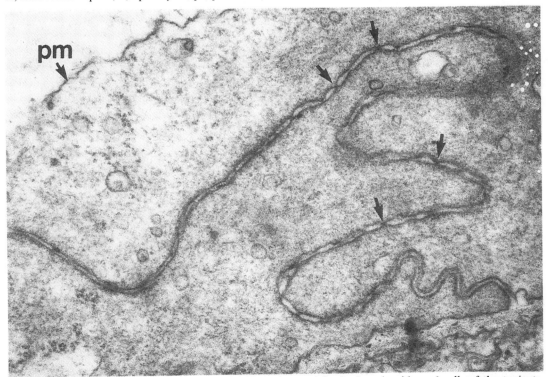

Fig. 3 – The distribution of punctate tight junctions (arrows) in the larval epidermal cells of the tunicate, *Sidnym elegans*. × 54 000. (Dr D. Georges, University of Grenoble.)   pm, plasma membrane.

contain potent hydrolytic enzymes. They exist in a variety of forms, for example, secondary lysosomes contain enclosed structures undergoing intracellular digestion (Figs 58–64). Chloroplasts are organelles found in plant cells. They possess a complex multilayered membranous structure and are involved in photosynthesis (Figs 1bc, 175, 176). In addition, in both plant and animal cells, numerous vesicles, vacuoles and secretory products found in the cytoplasm are also surrounded by membranes (Figs 98–102).

A complex system of membranes pervades the ground cytoplasm of cells to form numerous compartments and subcompartments. The term vacuolar system is used to indicate that the cytoplasm is generally separated into two parts; one contained within the system and the other remaining outside – the cytoplasmic matrix proper. The nuclear envelope and Golgi complex belong to the vacuolar system, but the major part is formed by the endoplasmic reticulum. The nucleus comprises a porous nuclear envelope, chroma-

tin material and nucleolus (Figs 24–30). The Golgi apparatus consists of a variable number of closely apposed flattened saccules enclosing spaces called cisternae. This organelle is involved in forming protein or complex polysaccharide secretory products, which are discharged at the cell surface as secretory granules (Figs 42–46). The endoplasmic reticulum (ER) may be differentiated into a granular or rough ER which consists of many membranous stacks containing ribosomes, which enclose spaces called cisternae, and an agranular or smooth ER composed of branched and anastomosing tubules (Figs 31–41).

In spite of the complex structural organization of organelles, important constituents of the cytoplasm are found in the matrix (ground cytoplasm) which lies outside the vacuolar system. The matrix constitutes the true internal milieu of a cell and contains glycogen particles, soluble enzymes, the ribosomes (polyribosomes) the principal structures of the machinery of protein synthesis, microtubules and

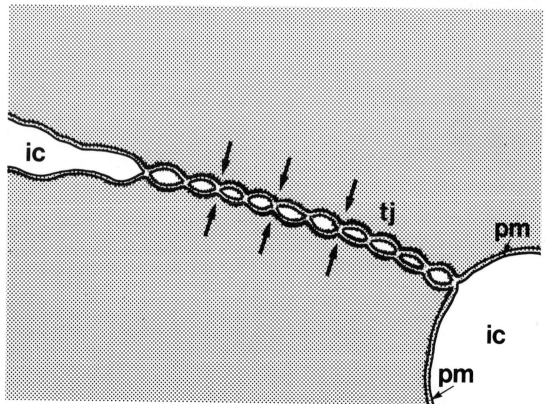

Fig. 4 – Diagram of punctate tight junctions (tj) illustrating the fusion of adjacent lateral membranes (arrows) and the absence of an intercellular space (ic).
pm, plasma membrane.

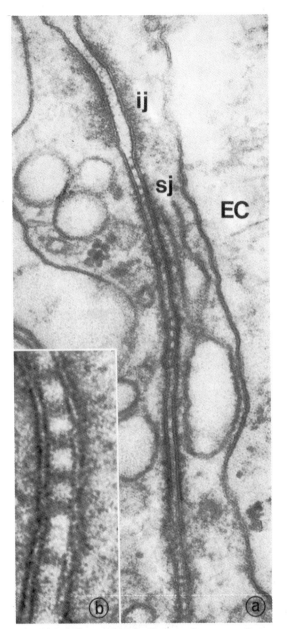

microfilaments (Figs 73–77). Microtubules maintain cell shape in a cytoskeletal capacity, and participate in intracellular movement by changing cell shape and the position of cell components. Similarly, microfilaments provide cells with rigidity as well as tensile strength and resilience. The cytoplasmic matrix is, in addition, the site of colloidal changes of the protoplasm and of the production of many cytoplasmic differentiations such as keratin and myofilaments (Figs 118, 119, 136).

PLASMA MEMBRANE

The plasma membrane, thoughout the life of a cell, controls the entrance and exit of molecules and ions. This function, of regulating the exchange between a cell and its external environment, enables a cell to maintain an internal milieu which is different from that of its external environment. The plasma membrane is composed mainly of protein, lipid and a small percentage of oligosaccharides that may be attached to either the lipids (glycoli-

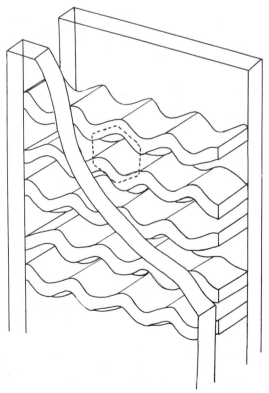

Fig. 5(a) – The septate desmosome between adjacent nephrocytes in *Achatina achatina*. The intermediate juncton (ij) bears microfilaments on its cytoplasmic surface, and the intermediate cellular space contains a finely filamentous material. In the septate junction (sj), the intercellular space is narrow and is crossed by septa. × 108 000. (Dr P.F. Newell, University of London.)

Fig. 5(b) – High-magnification electron micrograph showing that each septum lacks the unit-membrane structure of the adjacent plasma membranes. × 334 000. (Dr P.F. Newell.)

Fig. 5(c) – Diagram of a septate junction. (After, Gilula, Branton & Satir.)

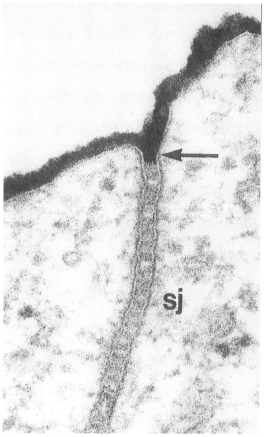

sj

*

Above

Fig. 6(a) – The septate junction in the spermatocyte envelope cells of the insect *Heliothis virescens* exposed to ruthenium red. The abrupt stoppage of the ruthenium red at the outer edge of the septate junction (arrow) is clear, as is the lack of any fusion, or even close apposition, of the membranes of adjacent cells in this location. × 194 000. (Dr K.M. Baldwin, Howard University.)

Above

Fig. 6(b) – Freeze-fracture replica of the septate junction between envelope cells. No fusion of membranes at the outer edge of the septate junction can be detected (*). × 133 000. (Dr K.M. Baldwin.)

Right

Fig. 7(a) – The septate junction between myoepithelial cells of the sea anemone, *Metridium senile*. A continuous series of 14 cross-sectioned septa is bracketed. Arrows indicate spherical densities adjacent to each septum. × 67 000. (Dr M.C. Holley, Bristol University.)

Far right above

Fig. 7(b) – High-magnification electron micrograph showing that each septum is composed of a filament core with pairs of lateral projections from each side (arrows). × 200 000. (Dr M.C. Holley.)

Far right below

Fig. 7(c) – Diagram of septa and adjacent cell membranes. Septa are composed of a filament core with a double row of lateral projections from each side. The cell membranes are corrugated. The black circles represent cross-sectioned filament cores, and the hatched circles represent electron-dense spheres on the cytoplasmic surfaces of the membranes. (Dr M.C. Holley.)

16

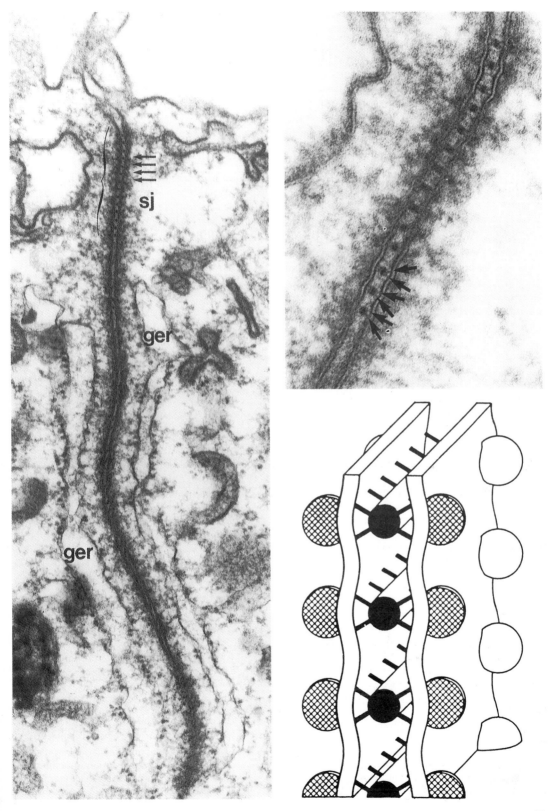

pids) or the proteins (glycoproteins). There is, however, a wide variation in the lipid : protein ratio between different cell membranes. The plasma membrane of most cell types appears to be three layered, i.e. two outer electron-dense layers about 2.0 nm thick and a middle electron-transparent layer about 3.5 nm thick (Fig 2, 15). This structure is called the 'unit membrane'. Protein is the main constituent of the outer electron-dense layers and lipid of the electron-transparent middle layer (Fig 2, 15).

INTERCELLULAR JUNCTIONS

In multicellular organisms, there are specialized regions of contact between the plasma membranes of adjacent cells. These specializations or intercellular junctions provide the structural means by which the groups of cells can adhere and interact. Among vertebrates and invertebrates three main functional categories of junction can be identified.

(1) Tight junctions.
(2) Gap junctions.
(3) Desmosomes.

These junctions are often arranged as pairs between adjacent cell membranes. A notable feature in invertebrates is the variety of forms each junction type can take.

(1) Tight junctions and septate junctions
In the vast majority of invertebrate tissue, no tight junctions have been located, instead, septate junctions occur in their place. These junctions have an occluding function, allowing concentration gradients to be established and maintained across an epithelium. They prevent large molecular substances from crossing epithelia via an intercellular route. In addition, they give firm cell adhesion and may play a role in intercellular communication.

Tight junctions (synonym: zonulae occludentes)
These occur at points where adjacent plasma membranes fuse to become a single membrane structure. Thus the total width of a tight junction is equal to that of the component membranes. Tight junctions are the most luminal elements in vertebrate junctional complexes, where they form belt-like regions between the plasma membranes of adjacent

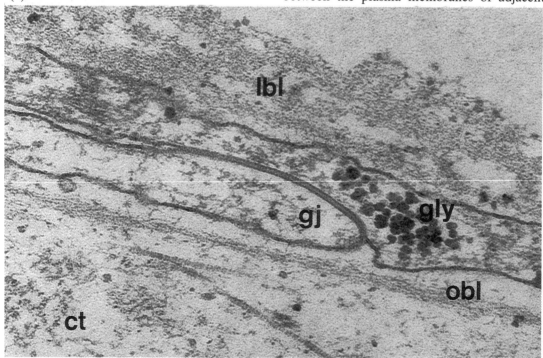

Fig. 8 – Overlapping cytoplasmic processes from adjacent peduncular myoepithelial cells in the blood vessel of the polychaete *Pomatoceros lamarckii*. There is a gap junction (gj) between the two apposed membranes. A thick multilayered luminal basal lamina (lbl) lines the distal luminal surface, whereas a single-layered outer basal lamina (obl) borders the proximal surface of the cells. × 120 000. (Dr A. Bubel.)

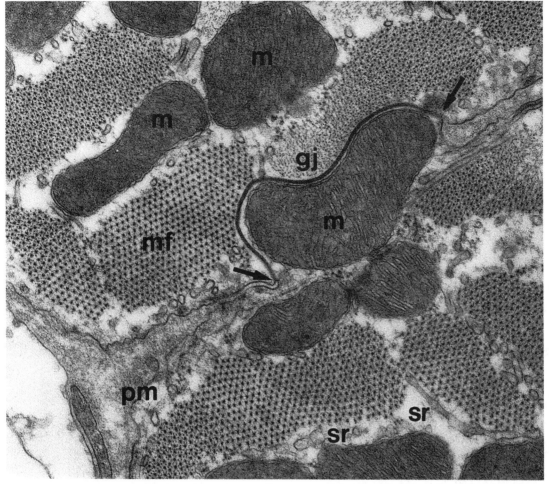

Fig. 9(a) – Transverse section through two mammalian myocardial cells. The apposed sacrolemmas form a gap junction (between arrows) which conforms closely to the profile of a mitochondrion (m). × 39 000. (Dr M.S. Forbes, University of Virginia.)

cells. In contrast, in invertebrates they are seen as a series of punctate points between adjacent plasma membranes (Figs 3, 4).

Septate junctions   In most invertebrate tissues, septate junctions form belts around the apical ends of cells lining luminal spaces and external bounding epithelia. They usually occur together with belt desmosomes. The belt desmosomes lie nearest to the free surface of cells, and the septate junctions lie immediately basal to these. (Fig 5a, b). However, septate junctions are found predominantly above any gap junctions or spot desmosomes in epithelia. In septate junctions, the intercellular space (between 15 nm and 18 nm wide) is traversed by a series of septa. The septa are long, thin-

walled structures running between cells, roughly parallel to cell apices. They tend to be more closely and regularly spaced towards the apical surface. The favoured role for the septate junction is that of an occlusion barrier. Indirect support for this comes from the observations that the epidermis of marine organisms typically have fewer septa than freshwater or terrestrial organisms. This might be expected as a result of reduced concentration gradients across the epithelial layer. However, in the insect, *Heliothis virescens,* at the outer edge of septate junctions, which join spermatocyte envelope cells, a novel occluding structure is found. The barrier is not simply the septate junction but a ridge/groove similar to a single-stranded vertebrate tight junction.

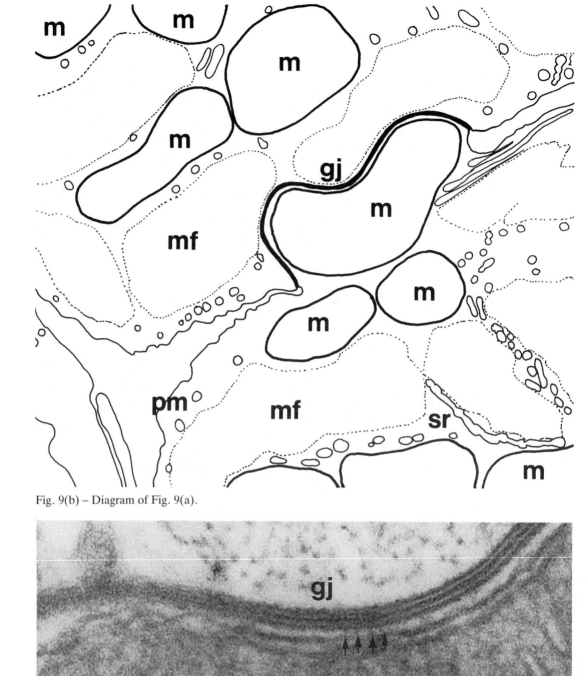

Fig. 9(b) – Diagram of Fig. 9(a).

Fig. 9(c) – High-magnification electron micrograph showing details of the opposition between the gap junction (gj) and mitochondrion (m). In one region, the surfaces of the two structures approach one another closely (separation approximately 8 nm) and they appear to be connected by strands of electron-opaque material (arrows). × 253 000. (Dr M.S. Forbes.)

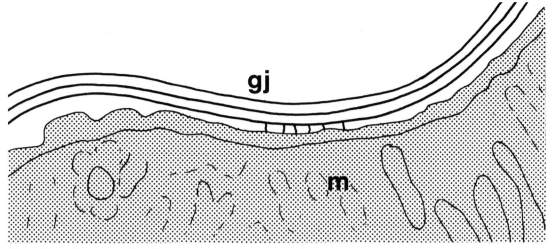

Fig. 9(d) – Diagram of Fig. 9(c).

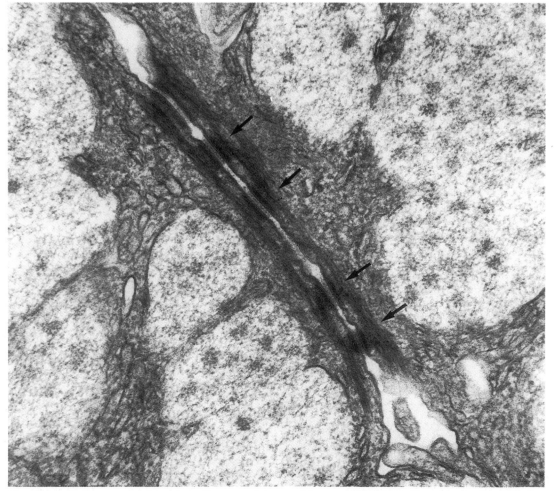

Fig. 10 – Spot desmosomes (maculae adhaerentes) (arrows) between the submaxillary gland cells of a rat. × 85 000. (Dr T. Jenkins, Portsmouth Polytechnic.)

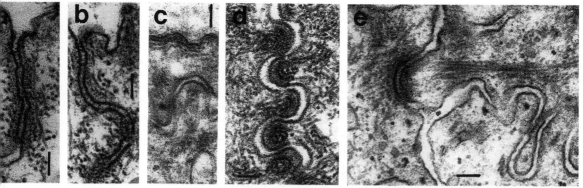

Fig. 11 – A hypothetical sequence, in the cephalochordate *Branchiostoma lanceolatum*, of the transformation of an intermediate junction (fascia adhaerens) to a desmosome-like formation by folding of the membrane is indicated by the sequence of (a)–(d). An electron-dense material is attached to both of the opposing membranes and bundles of intermediate filaments project into the cytoplasm (e). The main difference between these structures and true desmosomes is the lack of a distinctly structured, intracellular material. × 60 000. (Prof. Dr J. Bereiter-Hahn, University of Johann Wolfgang Goethe.)

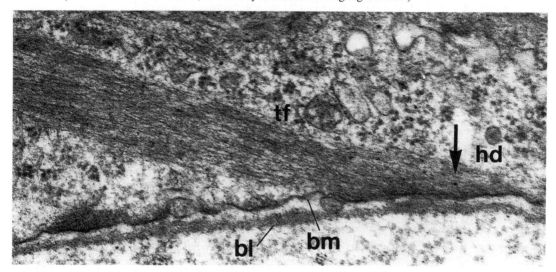

Fig. 12 – Mouse hair-follicle sheath cells showing the association of microfilament bundles (tf) with the basal plasma membrane (bm). The microfilament bundles show local concentration and terminate in dense plaques (hemidesmosomes, hd) on the basal membrane (arrow). There is filamentous or amorphous material between the plasma membrane and the basal lamina (bl). × 50 000. (Dr. S. Sugiyama, Sapporo Medical College.)

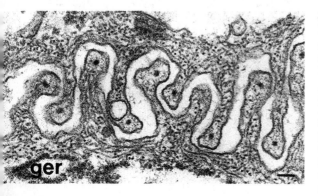

Left
Fig. 13 – The epidermal cells of *Branchiostoma lanceolatum* are linked by interdigitating ridge-like membrane folds. × 46 000. (Prof. Dr J. Bereiter-Hahn.)

Right
Fig. 14(a) – Rat duodenum intestinal cells. The microvilli (mv) are finger-like extensions of the plasma membrane (pm); each microvillus contains a core of filaments (f) that merge into a zone of filaments and amorphous material, the terminal web (tw), in the cytoplasm below the microvilli. × 68 000. (Dr T. Jenkins.)

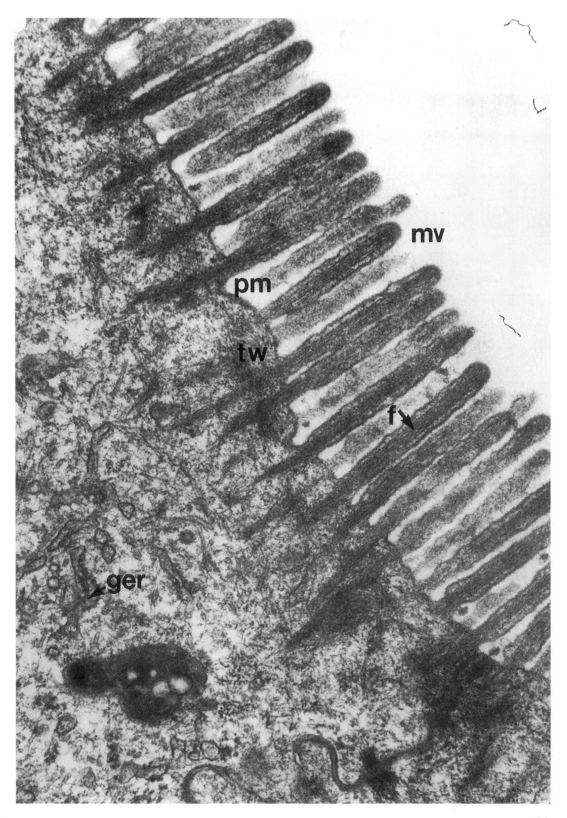

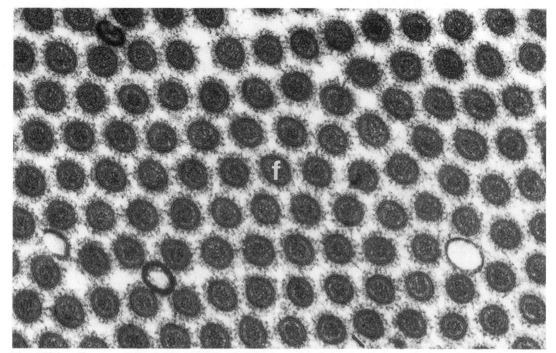

Fig. 14(b) – Cross-section of microvilli bordering the epithelial cells of the gut of the nematode, *Ascaris sp.* The microvilli are uniform in size and shape and are regularly packed. A core of filaments (f) is present in each microvillus, sectioned transversely. × 68 000. (Dr T. Jenkins.)

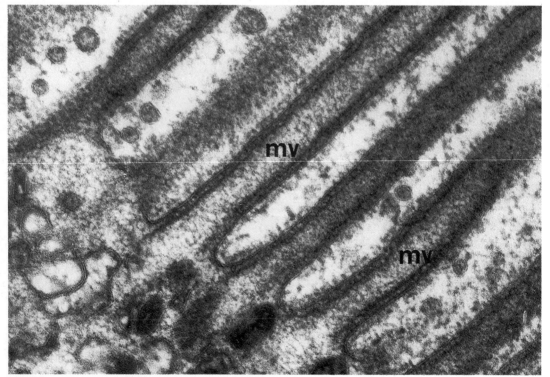

Fig. 15 – Microtriches (microvilli, mv) in the tegument of the adult tapeworm *Schistocephalus solidus.* × 70 000. (Dr G. Charles, Portsmouth Polytechnic.)

24

However, unlike a tight junction, no fusion or even close apposition of adjacent plasma membranes is evident in this region (Fig 6a, b).

In invertebrates, depending on the phylum and tissue, there is a wide variation in the type of septate junction present. However, the epidermal tissues of most have one or other of two types of pleated septate junction. The septa have an 18–23 nm periodicity pleating, with either prominent pegs at the apices of the pleats or less prominent pegs. Interestingly, in some situations where three (or more) cells come together, the septate junctions are modified to form tricellular junctions. The septate junctions from mesenteric myoepithelial cells and endothelial cells are described as double-septate junctions. This has been interpreted by some authors as a filament core 9.5–10.2 nm in diameter, with a double row of lateral projections along each side, which link the filament core to the adjacent membranes (Fig 7a–c).

(2) Gap junctions (synonyms: nexus; maculae communicantes)

Gap junctions appear as short stretches where apposing plasma membranes are separated by a distance of about 2-4 nm. There are hexagonal arrays of minute polygonal subunits in the intercellular space, and within each apposed plasma membrane. The overall thickness of the junction varies between 13 nm and 19 nm. The close apposition of plasma membranes in gap junctions may result in their being mistaken for tight junctions. Gap junctions are regions of electrical resistance for the cell-to-cell propagation of impulses in the form of ions. They are also the regions where molecules are exchanged between cells. Gap junctions occur in both vertebrate and invertebrate epithelia, muscle and nervous tissue (Figs 8, 9a–d).

(3) Desmosomes

These junctions are commonly found in both vertebrate and invertebrate tissues, where they provide points for cell-to-cell adhesion and cell-to-substrate adhesion. Desmosomes are, therefore, especially numerous in tissues subjected to mechanical stress, such as muscle-attachment sites, or in epithelia undergoing large-scale deformation, such epidermal tissues. There are three main types of desmosomes, namely, spot desmosomes, belt desmosomes and hemidesmosomes.

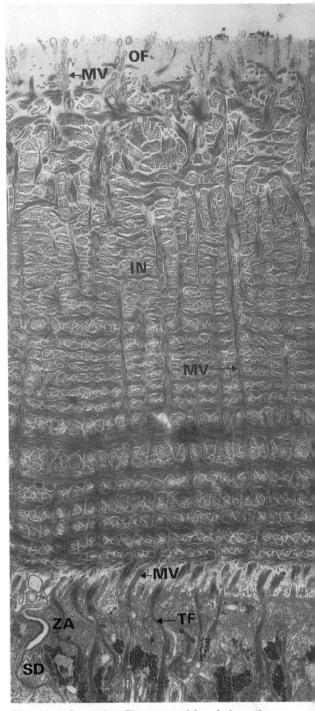

Fig. 16 – Opercular filament cuticle of the polychaete, *Pomatoceros lemarckii* is composed of an outer filamentous layer (OF) and inner component (IN) composed of orthogonally arranged layers of fibres. The cuticle is transversed by long microvilli (MV). × 7 600. (Dr A. Bubel.)

Spot desmosomes (Synonym: maculae adhaerentes) These occur in most instances as paired oval/disc-shaped structures 0.2 μm to 0.5 μm in diameter with an intercellular space of between 22 and 35 nm wide filled with electron-dense material, often containing laminar densities. Closely applied to the cytoplasmic surfaces of the junctional membranes are electron-dense plaques, which also show some continuity across the intercellular space. These plaques provide an area of anchorage for converging filaments and may structurally connect cells in series. Spot desmosomes allow the passive distribution of shear forces from each cell to tissue as a whole, and also protect plasma membranes from excessive deformation. They are found in vertebrate and invertebrate tissues that are subjected to large volume changes, e.g. oesophagus, bladder, epidermis of blood sucking bugs (Fig 10).

Belt desmosomes (synonyms: zonulae adhaerentes; fasciae adhaerentes; intermediate junctions) These junctions are superficially similar to spot desmosomes. However, they differ in that they form a belt around the extremities of the lateral borders of epithelia. In invertebrate tissue they are usually the outermost junction occuring above the septate junction, when present. The intercellular space

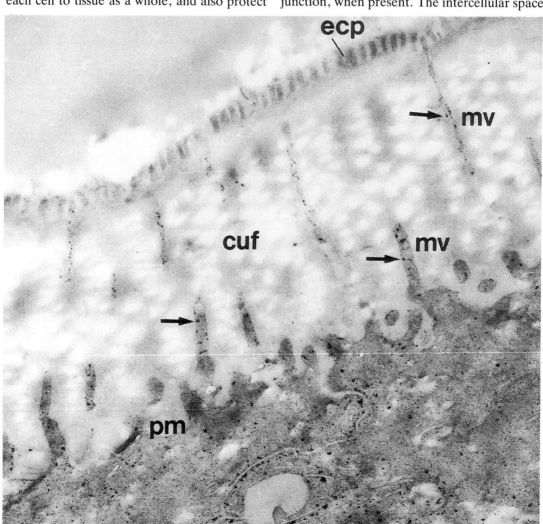

Fig. 17 – The location of acid phosphatase activity in the epidermal cells of the oligochaete *Eisenia foetida*. Reaction products (arrows) are located in relation to the ascending microvilli (mv) and are rare at the level of the apical plasma membrane (pm) between mirovilli and absent from epicuticular projections (ecp). × 33 000. (Dr K.S. Richards, University of Keele.)   cuf, cuticular collagen fibres (electron-lucent).

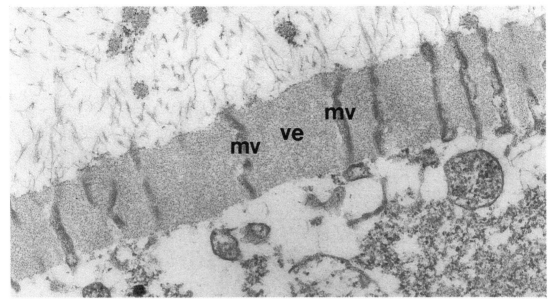

Fig. 18 – The oolemma of the bivalve *Mytilus edulis* egg is amplified by the formation of numerous microvilli (mv). Associated with the microvilli is a fuzzy coat that comprises the vitelline envelope (ve). × 30 000. (Dr A. Bubel.)

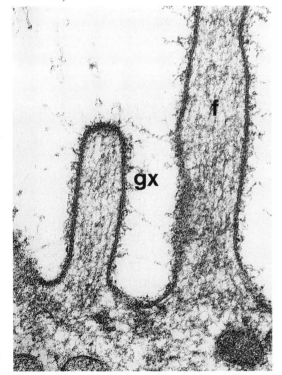

Fig. 19(a) – The microvilli of guinea-pig yolk-sac epithelial cells showing well-delineated plasma membrane and glycocalyx (gx); a core of filaments (f) is discernible. × 88 000. (Dr B.F. King, University of California.)

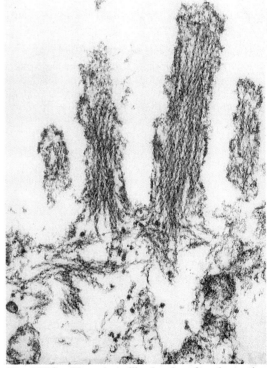

Fig. 19(b) – The microvillus border after decoration with myosin SI subfragments, indicating actin filaments in the cores of microvilli and in short rootlets, and the interweaving of the short rootlets with decorated filaments running more or less parallel to the cell surface. × 64 000. (Dr B.F. King.)

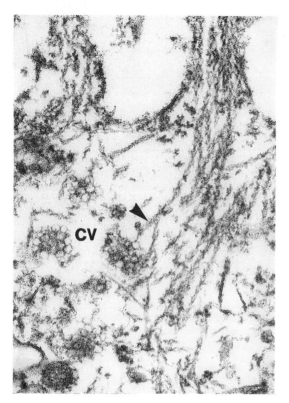

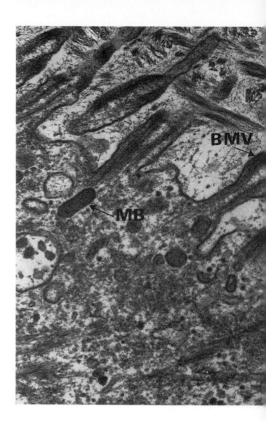

Above
Fig. 19(c) – The relationship between the basal end of a microvillus and coated vesicles (cv). The clathrin basket of the coated vesicles is well stained. An actin filament emanating from the microvillus appear to impinge on one of the coated vesicles (arrow). × 94 000. (Dr B.F. King.)

Above right
Fig. 20(a) – Epidermal cells of the polychaeate, *Pomatoceros lamarckii* with basal microvilli distensions (BMV) containing filaments (TF) and slender microvilli (MV) which penetrate the cuticle. × 28 000. (Dr A. Bubel.)

Above far right
Fig. 20(b) – Diagram of Fig. 20(a).

Below right
Fig. 20(c) – Cross-section through the distended bases of *P. lamarckii* epidermal cells revealing bundles of filaments (TF) in their cores. × 49 000. (Dr A. Bubel.)

Below far right
Fig. 20(d) – Diagram of Fig. 20(c).

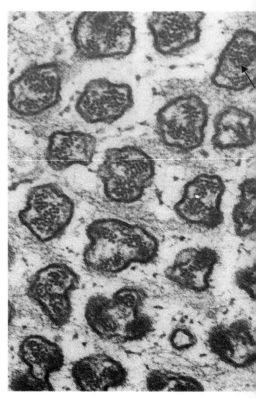

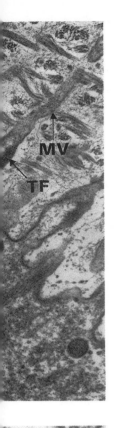

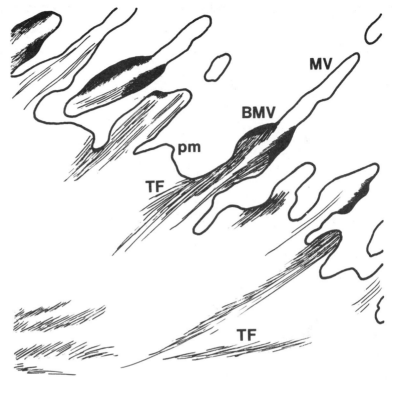

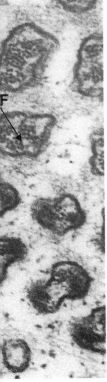

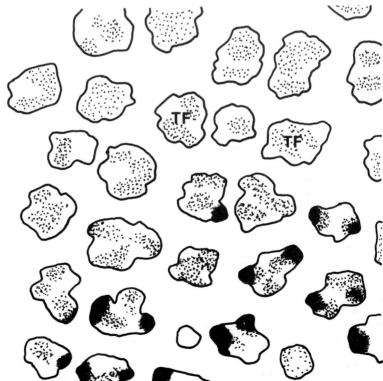

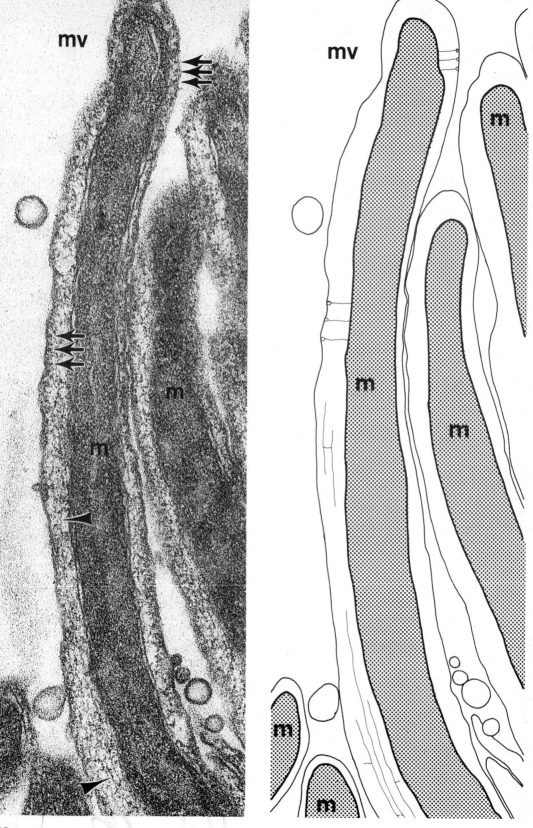

30

is on average about 20 nm wide and filled with a medium electron-dense amorphous/filamentous material, in which a central line of higher density is occasionally present. At the level of the belt desmosome, the adjacent cytoplasm lacks the compact plaques of spot desmosomes, but is filled with filaments which form part of the terminal web. These junctions are commonly found between the lateral membranes of columnar epithelial cells, especially those undergoing contraction and relaxation, e.g. epidermis immediately below the cuticle of earthworms, sea anemone tentacles and the intercalated discs of cardiac muscle cells (Figs 5, 11a-d).

Hemidesmosomes These are literally half desmosomes that commonly occur along the basal plasma membrane of stress-bearing epithelial cells. Hemidesmosomes are found anchoring epidermal cells to a basement membrane, to an extracellular collagenous matrix, or to other structures, e.g. muscle-extracellular (skeleton, cuticle) attachment sites. In such cells, tonofilaments pass through the cytoplasm and terminate in an electron-dense plaque. A network of filamentous material on the external plasma membrane may then link the hemidesmosomes to a basement membrane (basal lamina) or to a collagenous matrix (Fig 12).

In addition to intercellular junctions, adjacent cells may be held together, to varying degrees in different regions of a tissue, by the interdigitation of lateral plasma membranes (Fig 13). The presence of simple interdigitations between, for example, epidermal cells may allow stretching of the epidermis.

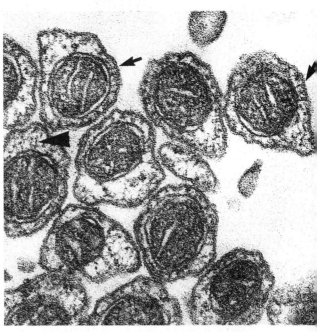

Fig. 21(c) – A cross-section in the region of the tips of large microvilli of the primary cells. In some areas knob-like portosomes appear to bridge the gap between the plasma membrane and the outer mitochondrial membrane (arrows). Bridges from the filaments to the outer mitochondrial membrane are also evident (arrow heads). × 100 000. (Dr T.J. Bradley.)

Far left
Fig. 21(a) – Longitudinal section of large microvilli (mv) in the primary cell of the Malphigian tubules of the mosquito *Aedes taeniorhynchus*. Each microvillus contains a mitochondrion (m) and narrows at the top so that the mitochondrial outer membrane and the plasma membrane lie much closer together in this region. Longitudinal filaments (arrow heads) run between the plasma membrane and the outer mitochondrial membrane. Knob-like structures are also apparent on the plasma membrane (arrows). × 80 000. (Dr T.J. Bradley, University of California.)

Left
Fig. 21(b) – Diagram of Fig. 21(a).

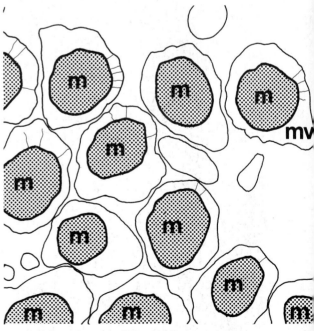

Fig. 21(d) – Diagram of Fig. 21(c).

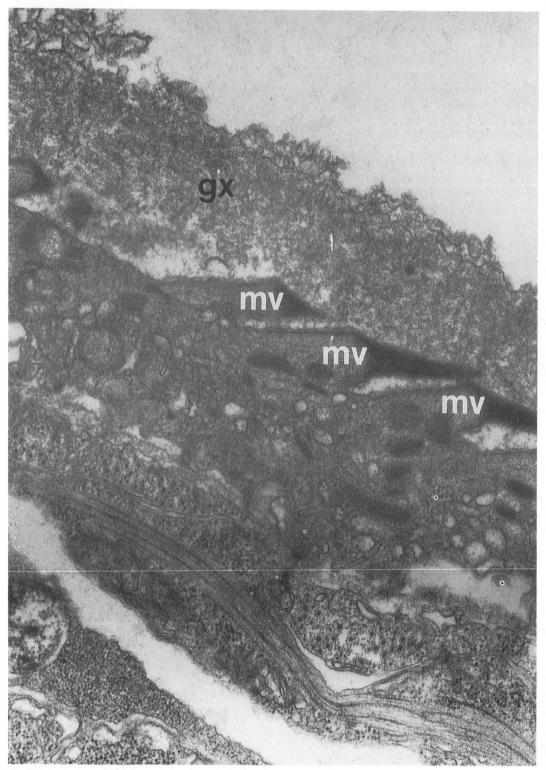

Fig. 22(a) – Developing microtriches (microvilli, mv) in the tegument of an eight day-old plerocercoid larva of the tapeworm, *Schistocephalus solidus*. × 78 000. (Dr G. Charles.)   gx, glycocalyx.

MICROVILLI

Projecting from the surface of most cells are a varied number of differently shaped cellular extensions, collectively known as microvilli. In epithelial cells, microvilli increase cell surface and are available for exchange processes. However, in wandering cells, such as macrophages and lymphocytes, they may play a role in cell movement.

Generally, microvilli do not maintain a constant height and diameter. On the luminal surface of absorbtive intestinal epithelial cells, microvilli form a highly organized array of straight, equal-sized cellular extensions, sometimes referred to as a striated border (Figs 14, 15). Long, closely packed microvilli form the rhabdome of the photoreceptor cells in some polychaetes also (Fig 96). However, the absorptive cells of the proximal tubules of the kidney possess a less-ordered arrangement of long and densely packed microvilli of varying dimensions, called a brush border. In contrast, the absorptive cells of the yolk-sac epithelium possess dome-shaped apices that result in the spreading apart of the microvilli. Invertebrate epidermal cells frequently bear long microvilli that penetrate a cuticle (Fig 16). In such microvilli the activity of the enzyme acid phosphatase has been located, which suggests that they may be involved in the uptake of substances from the surrounding environment (Fig 17). The oocytes of both vertebrates and invertebrates possess long microvilli that penetrate an acellular vitelline envelope in order to facilitate absorption (Fig 18).

Microvilli are usually devoid of cell organelles, except for a varied number of filaments which form a central core. In most microvilli the central core is made up of between five and forty such filaments (Figs 19a–c). The microvilli of invertebrate epidermal cells, such as those of some polychaetes, possess swollen, irregularly outlined bases containing filaments with distal-branched circular components, devoid of filaments, which end in terminal swellings (Figs 20a–d, 22b). Similar large specialized branched microvilli serve to attach so-called invertebrate tendon cells to exoskeletal structures, such as a fibrous cuticular flange in some polychaetes. In contrast, ion-transporting epithelial cells lining the Malpighian tubules of the mosquito possess both long (3–4 μm in length) and short microvilli, which contain filaments, and in addition, mitochondria and some extensions of the ER (Fig 21). The protoplasmic surface of the plasma membrane of both types of microvilli is coated with short knobs (portasomes) about 25 nm long and 5 nm in diameter, which are believed to play a role in ion transport. Microvilli may act as membrane stores that contribute to the

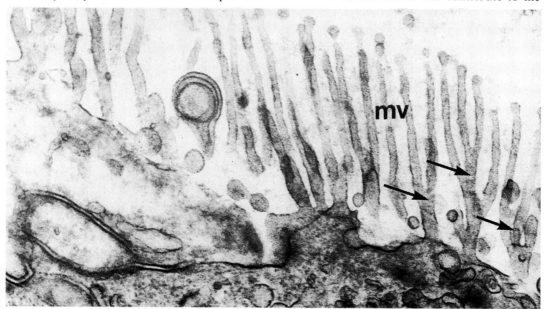

Fig. 22(b) – Branched microvilli (arrows) on branchial filament epithelial cells of the polychaete *Spirorbis spirorbis*. × 9 500. (Dr C.A.L. Fitzsimons, Portsmouth Polytechnic.)

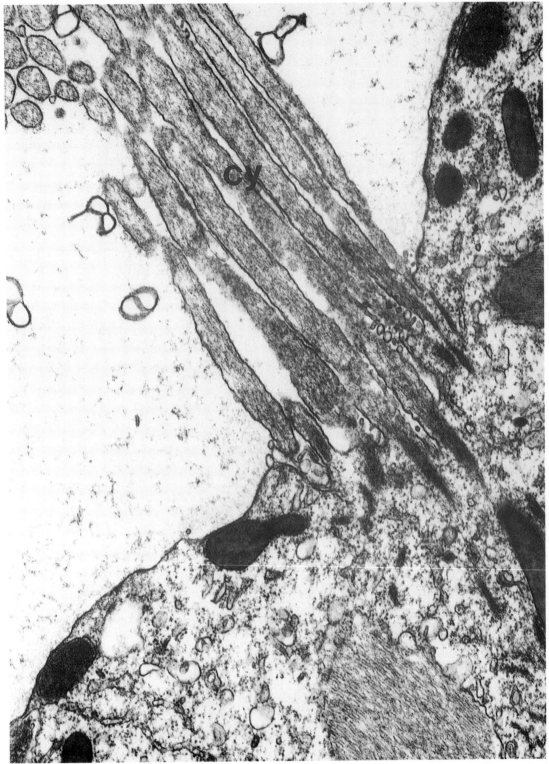

Fig. 23(a) – A longitudinal section of a tuft of cytospines (cy) projecting from the surface of a large oocyte of the sea anemone, *Actinia fragacea*. × 24 000. (Dr A. Larkman, Portsmouth Polytechnic.)

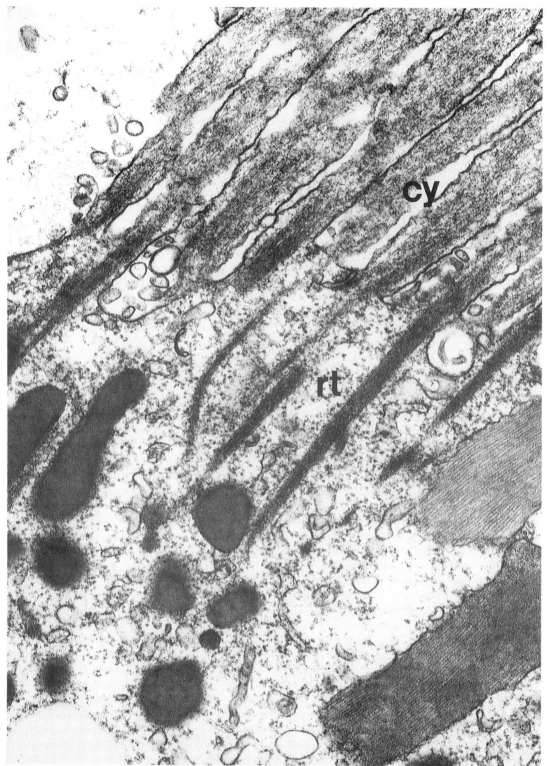

Fig. 23(b) – Basal part of a tuft of cytospines sectioned longitudinally, showing the fibrillar rootlets (rt) extending from their bases. × 85 000. (Dr A. Larkman.)

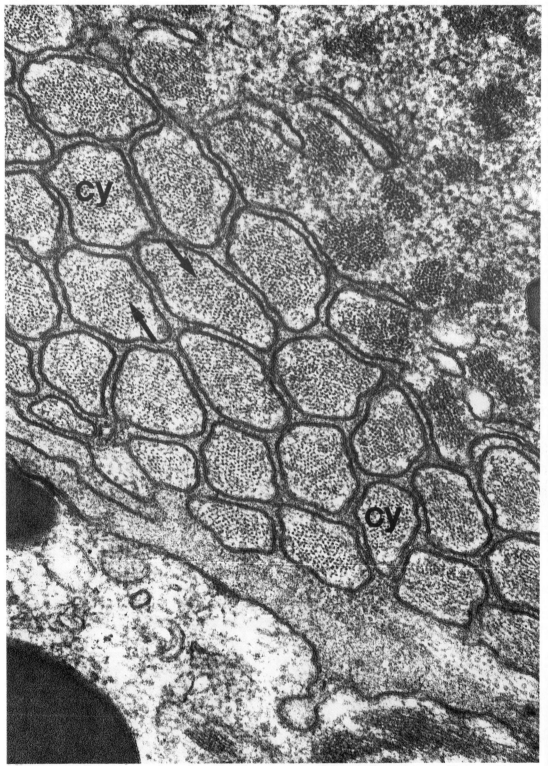

Fig. 23(c) – A part of a cytospine tuft sectioned transversely. The core filaments are evident, and in some cases they appear to be aligned in straight rows (arrows). × 12 000. (Dr A. Larkman.)

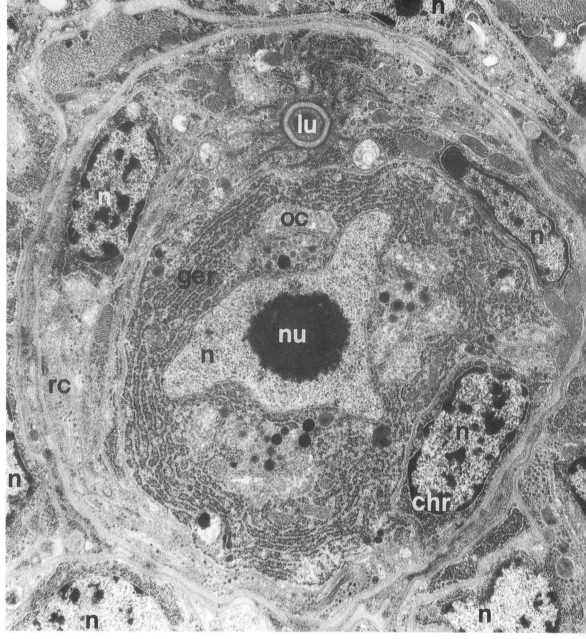

Fig 24(a) – The nucleus (n) of the active odontostyle cell (oc) of the nematode, *Xiphinema americanum* is large and lobed with a large nucleolus (nu), and contains little condensed chromatin (chr). Radial cells (rc) possess ovoid to rod-shaped nuclei which contain condensed chromatin. × 17 000. (Dr R.F. Carter, Ontario Veterinary College.)
lu, lumen of the oesophagus.

extra cell surface required as the cells increase in diameter. Number and distribution density of microvilli decrease in sea anemone myoepithelial cells when cell diameter increases.

The plasma membranes of most microvilli are covered by an extracellular coat or glyco-calyx, which is composed of fine filaments. The thickness of the glycocalyx is extremely variable, ranging from 7.5 to 20 nm in vertebrate absorptive cells and from 17 to 800 nm in invertebrate epidermal cells. In most cases, the glycocalyx contains neutral and acidic muco-polysaccharides (Figs 19a, 22a).

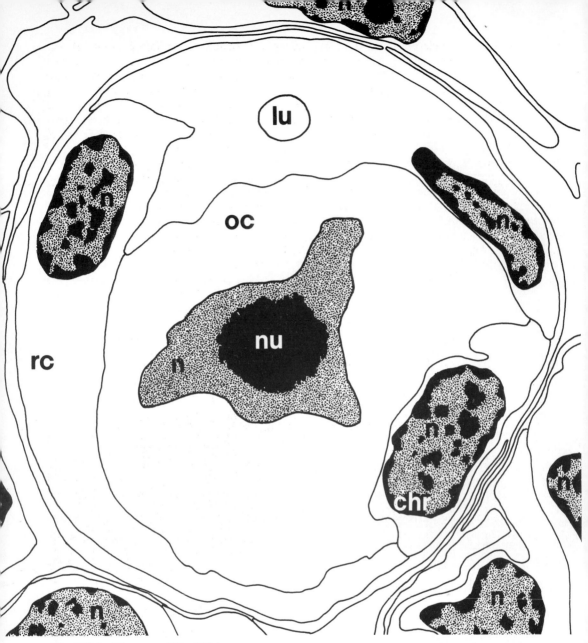

Fig. 24(b) – Diagram of Fig. 24(a).

CYTOSPINES

These are large versions of microvilli found on the eggs of many animal groups. They are large finger-like extensions of the apical plasma membrane, ranging from 1 to 25 μm in length and from 0.1 to 0.4 μm in diameter, and are generally packed together. The lumen of each contains an axial core of 20–300 parallel filaments, each about 5 nm in diameter and containing actin. The filaments extend into the cytoplasm from the base of a cytospine and link up with the terminal web. The filaments may continue from the bases of cytospines to rootlets, which may extend for up to 1.5 μm into the cytoplasm (Fig 23a–c).

NUCLEUS

A nucleus is present in all cells except in the blood cells, erythrocytes and platelets. Most cells have only one nucleus, whereas some contain two (binucleate), e.g. liver parenchyma cells, and some contain many, e.g. skeletal muscle cells. The shape of the nuclei varies from round to ovoid, rod-shaped or cup-shaped (Figs 24, 153b). They may also possess several indentations and in some instances be lobed (Figs 24, 153c). A nucleus contains a nucleolus, composed principally of ribonucleoprotein and chromatin, which contains deoxyribonucleoproteins. The nucleus is surrounded by a nuclear envelope (nuclear membrane) which has an important role in molecular exchanges between the nucleus and cytoplasm. The nucleus contains genetic information and has a major influence on the metabolic activities of the cytoplasm.

The nuclear envelope is a membranous sac consisting of (a) an outer nuclear membrane, (b) an inner nuclear membrane, (c) a perinuclear cisterna, and (d) nuclear pores. The nuclear envelope is interpreted as a flattened cisterna of the ER having ribosomes only on its outer surface. The outer and inner nuclear membranes are about 8 nm thick and have a trilaminar structure similar to that of the plasma membrane. In most cells, the cytoplasmic face of the outer membrane is studded with ribosomes, whereas the inner membrane may be lined by an electron-dense material. The perinuclear cisterna may occasionally communicate with the cisternae of the ER in the cytoplasm. The nuclear envelope is perforated at certain points by nuclear pores. Around the margins of the nuclear pore, the outer and inner nuclear membranes are continuous and form the annulus (ring) of the pore (Fig 25a, b). A thin diffuse diaphragm and/or plug-like structure stretches across or fills the pore. The number of pores varies from 40 to 45 per μm in nuclei of various plant and animal cells. At their highest density, the pores are packed in a hexagonal array with centre-to-centre distances of about 150 nm. In mammals nuclear pores account for at least 10% of the surface area. Nuclear pores are considered potential channels for exchange and interaction between the karyoplasm (nucleus) and the cytoplasm. They are also considered to act as a barrier between the karyoplasm and the cytoplasm.

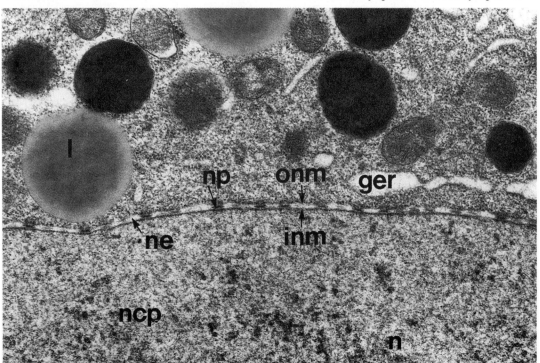

Fig. 25(a) – Section through part of the nucleus of an early vitellogenic oocyte of *Mytilus edulis*. The nucleoplasm (ncp) is of a low density and has a flocculent appearance. The nuclear pores (np) are numerous. × 30 800. (Dr A. Bubel.)
inm, inner nuclear membrane; l, lipid; onm, outer nuclear membrane.

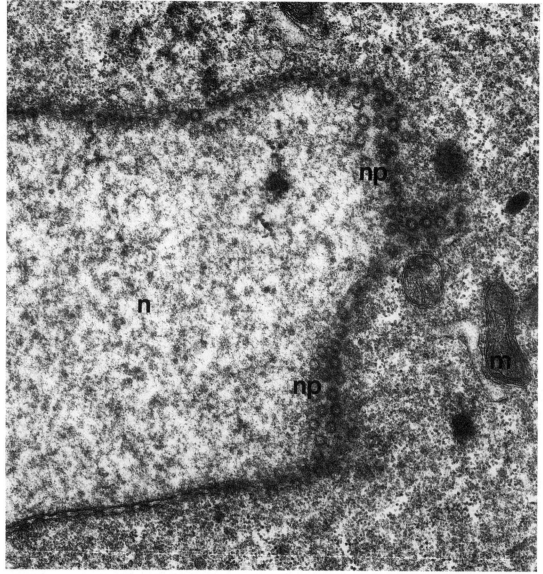

Fig. 25(b) – Tangential section through part of an early vitellogenic oocyte of the bivalve *Mytilus edulis*. Nuclear pores (np) are very numerous. × 30 800. (Dr A. Bubel.)

The genetic substance of the nucleus is referred to as chromatin material. In the karyoplasm, the chromatin material may be present in a condensed state, i.e. heterochromatin, or in a more dispersed form, i.e. euchromatin (Fig 24). The euchromatin occupies the light areas of the karyoplasm, and the heterochromatin constitutes the irregular, dense, coarsely granular areas. The heterochromatin material forms a network of electron-dense non-membrane-bound chromatin granules, i.e. karysomes. It is often distri-buted along the inner surface of the nuclear envelope, usually leaving the nuclear pores free. Some dense heterochromatin is attached to the nucleolus. In the interphase, as well as the mitotic nucleus, the genetic material is present in the form of chromosomes. It is presently believed that the heterochromatin part of the nucleus represents the contracted (coiled, non-dispersed) part of the chromosomes. The euchromatin is, in contrast, assumed to contain the extended (uncoiled, dispersed) part of the chromosome. The

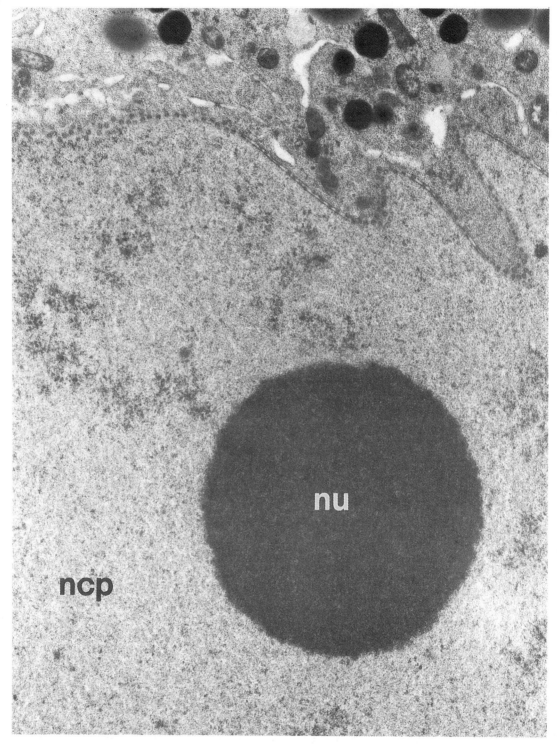

Fig. 26 – Part of the nucleus of an oocyte of the bivalve *M. edulis*. The nucleoplasm (ncp) is of a low density and the nucleolus (nu) is discrete and not associated with other structures. × 9 400. (Dr A. Bubel.)

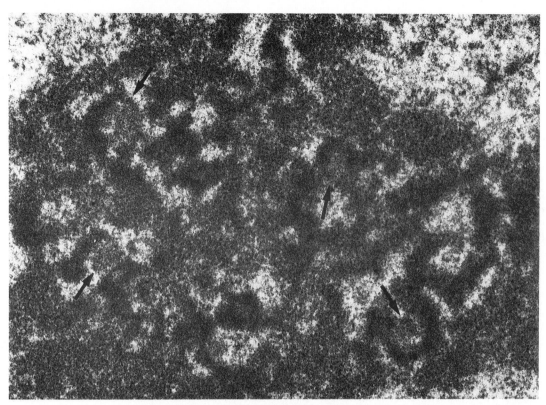

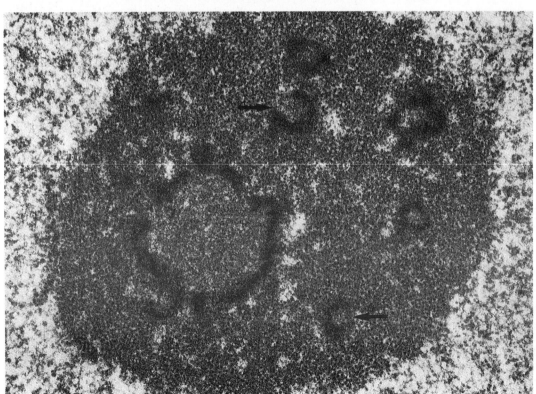

42

euchromatin areas of the nucleus are also believed to contain a relatively high proportion of nuclear sap, the fluid component of the karyoplasm (nucleoplasm).

An interphase nucleus contains one or several nucleoli. Nucleoli are discrete, round or ovoid non-membrane-bound bodies. They occur freely in the karyoplasm, or are attached to the inner nuclear membrane and are particularly large and prominent in rapidly growing cells and cells with a high rate of protein synthesis, i.e. undifferentiated cells (Figs 24, 26). However, during cell division, the nucleolus disappears. The nucleolus consists of (a) a densely coiled fibril, the nucleolonema (pars densa), which is composed of small particles 10 nm to 15 nm in diameter and short delicate filaments 5–7nm in diameter, believed to contain nuclear RNA and non-ribosomal nucleolar proteins respectively; (b) the pars amorpha (pars fibrosa) which occupies irregular and spherical spaces within the coiled nucleolonema and contains a finely filamentous substance and some chromatin particles; and (c) the condensed boundary (nucleolus-associated chromatin) which contains the so-called nucleolus-associated chromatin (Fig 27a, b).

During cell division in some cells, microtubule bundles are present between the mitotic apparatus (Fig 28a–c). In addition, the nucleolus disappears and its content is disrupted and attached to chromatids, which act as nuclear organizer material. Each nucleolus is in contact with a pair of chromosomes;

the point of union is a special region called the nuclear organizer (Fig 29a–d). During meiosis, nuclei are found to contain synaptinemal complexes, which are thought to represent structures by which homologous chromosomes become aligned and paired (Fig 29a). In the nuclei of many cells, there are three-layered synaptinemal complexes. These consist of two outer dense granular areas lying either side of a less-dense, more finely granular, central lamina. The outer areas are separated from the central lamina by a gap of some 20 nm.

Structures known as nuage have been reported from germ cells of a great many animal species, ranging from coelenterates to mammals. Nuage material in oocytes appears as conjoined cylinders of electron-dense material. Each cylinder may contain a cylindrical core made up of fine fibrils arranged longitudinally. Nuage material is also associated with small groups of membranous tubules (each tubule 80 nm in diameter). Where nucleoli and nuclear fibrillar bodies are found close to the nuclear envelope, areas of nuage material are found immediately outside it (Figs 30a, b). The association of nuage material with the nucleus is taken as evidence of nucleocytoplasmic exchange. It has been suggested that nuage material is formed in the nucleolus.

ENDOPLASMIC RETICULUM AND RIBOSOMES

An endoplasmic reticulum (ER) is present in the cytoplasm of most cells. It consists of a network of paired trilaminar membranes, on average 6 nm thick, which form vesicles, vacuoles and flattened cisternae. The ER is divided, on the basis of the presence or absence of attached ribosomes, into granular (rough-surfaced) ER and agranular (smooth-surfaced) ER, respectively. Ribosomes are minute angular particles, between 15 and 25 nm in diameter. They may be attached to the cytoplasmic aspect of ER membranes, either singly (monoribosomes) or in clusters as rosettes, spirals or helices (polyribosomes; polysomes). Either category of ribosomes may also occur free in the cytoplasm. The two types of ER and ribosomes are engaged in a great variety of functions, especially protein and steroid synthesis.

Left above
Fig. 27(a) – The nucleolus of a regenerating rat hepatocyte cell. In the nucleolus the characteristic nucleolenema-like structural organization of the ribonuclear proteins (RNP) is seen. A ribonuclear protein fibrillar component is evident at the periphery of fibrillar centres. Arrows indicate fibrillar centres. × 42 000. (Dr M. Derenzini.)

Left below
Fig. 27(b) – The nucleolus of a human tumour cell. In the nucleolus a large fibrillar centre is visible surrounded by a rim of ribonuclear protein (RNP) fibrillar structures. Other fibrillar centres are smaller in size (arrows). The ribonuclear protein components do not seem to give rise to a nucleolonema organization in the nucleolar body, where the ribonuclear protein granular components predominate. × 42 000. (Dr M. Derenzini.)

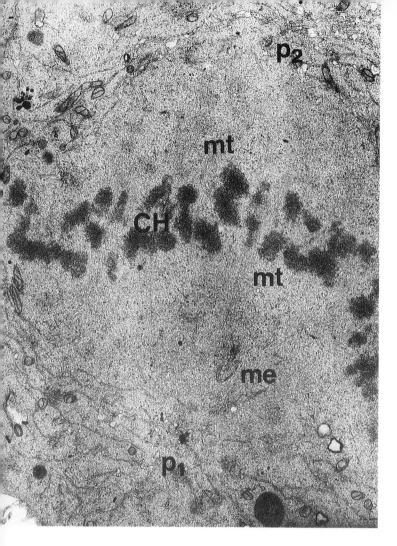

Above

Fig. 28(a) – The central region of a Hela cell in metaphase–early anaphase. The mitotic apparatus is surrounded by membranous cisternae. The two poles (p_1, p_2) are apparent; at p_2 a star-like arrangement of membranes is found. Within the mitotic apparatus bundles of microtubules (mt) and accumulations of membranes (me) are shown. × 9 600. (Dr N. Paweletz, German Cancer Research Centre.) CH, chromosomes.

Centre

Fig. 28(b) – Diagram of Fig. 28(a).

Right

Fig. 28(c) – The membranes (me) accompanying bundles of microtubules (mt) and chromosomes (CH) within the mitotic apparatus of a Hela tumour cell. × 31 500. (Dr N. Paweletz.)

Granular ER

It is generally accepted that both ribosome-studded granular ER and free ribosomes are major sites for the synthesis of proteins in a wide variety of cell types. In general, there is variability in the quantity and degree of organization of the granular ER in different cell types. The membrane patterns vary from one or two flat cisternae in most cells, to many widely distended cisternae in plasma cells, annelid gland cells, molluscan gland cells, odontostyle cells in nematodes and arthropod haemocytes (Figs 31b; 32a, 101a, d, 160), to stacks of widely distributed parallel flat membranes and cisternae in acinar cells of the pancreas and oocyte cells (Fig 31a). The cisternae usually contain a material of moderate electron density. In some cells, intracisternal granules (secretory granules)

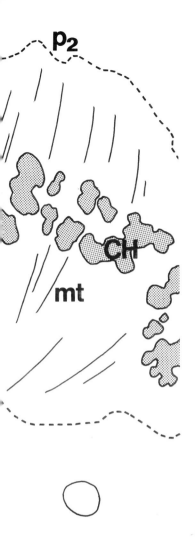
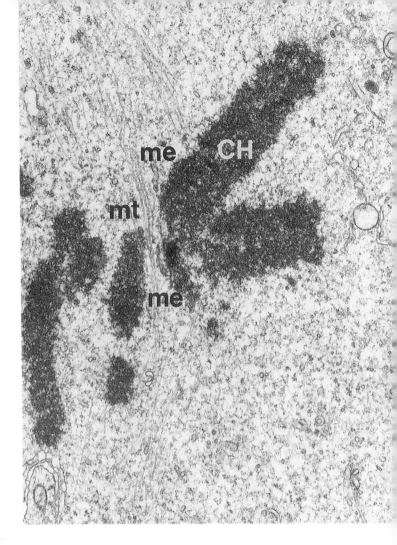

appear to develop in widely distended cisternae of the granular ER (Fig 32b). In such cells, there is little sign of Golgi complexes having a packaging function. Haemoglobin, haemocyanin and chlorocruorin molecules are found in the distended cisternae of granular ER in the pore and extravasal cells of molluscs and annelids, respectively (Fig 33a–d).

In contrast, the blood cell, cyanoblast, of the horseshoe crab is replete with free ribosomes and synthesizes the blood pigment haemocyanin. Helical polyribosomes (16–22) are found attached to the faces of growing crystals of haemocyanin (Fig 34). Granular ER profiles are often continuous with those of the agranular ER, of the Golgi zone and the nuclear membrane (Figs 39, 42, 101). Elements of the granular ER are also frequently associated with other cytoplasmic inclusions, such as mitochondria and lipid droplets (Fig 35). Frequently in cells, vesicles are observed to break away from the ends of the cisternae. These are referred to as transfer vesicles. Often the term microsome is associated with the granular ER. A microsome is a small, membrane-bound sphere, the outside of which is studded with ribosomes. However, microsomes are not present as discrete structures in the cell, but are broken up pieces of granular ER derived from damage caused to cells during analytical procedures.

(B) Agranular ER
The membranes of the agranular ER lack ribosomes. It is involved in a multiplicity of functions other than protein synthesis. There is an abundance of agranular ER in cells of the adrenal cortex (Fig 36), in interstitial cells of the testis, and in cells of the corpus luteum,

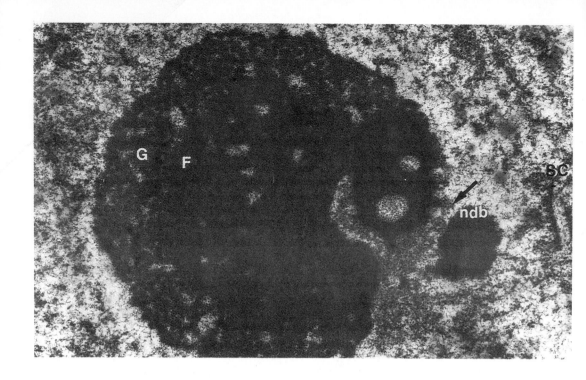

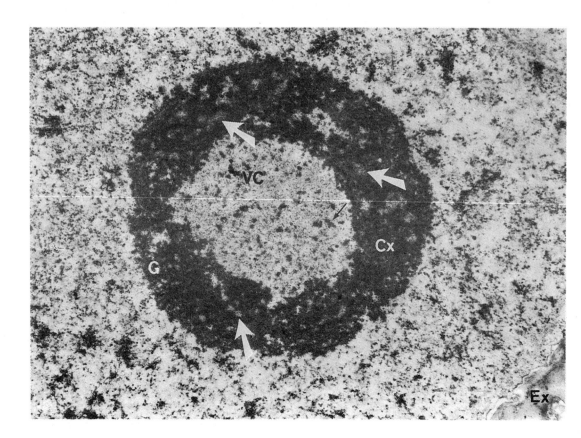

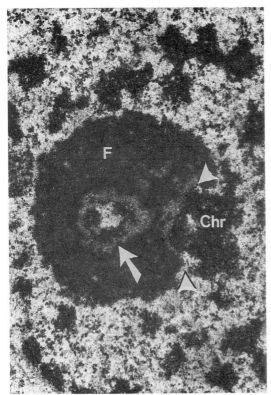

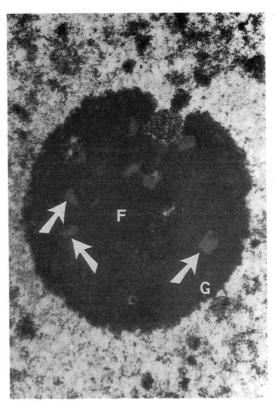

Fig. 29(c) – The generative cell of a bicellular pollen grain of *H. non-scriptus*. The nucleolus is exclusively fibrillar (F) in nature, and has large, heterogeneous fibrillar centres (arrows). Connections between the intranucleolar and extranucleolar are evident (arrow head) due to their large size. × 21 500. (Dr F.J. Medina.)

Fig. 29(d) – The vegetative cell of a bicellular pollen grain of *H. non-scriptus*. The nucleolus is made up of dense fibrillar (F) and granular (G) components. Small homogeneous fibrillar centres (arrows) are seen inside the fibrillar component. × 13 800. (Dr F.J. Medina.)

Left above
Fig. 29(a) – The ovaries of *Hyacinthoides non-scriptus* in the middle meiotic prophase I stage, with segregated nucleolus. The nucleolar organizer (NOR) (arrows) is present on the periphery of the nucleolar mass, associated with masses of condensed extranucleolar chromatin (arrow) and joined to nucleolar dense bodies (ndb) which are the products of its activity. The nucleolar organizer shows a homogeneous structure. × 26 000. (Dr F.J. Medina.)
F, nucleolar dense fibrillar component; G, nucleolar granular component; SC, synaptonemal complex.

Left below
Fig. 29(b) –*Hyacinthoides non-scriptus* microspore in the G2 interphase stage. A nucleolus with a large central vacuole (VC) with clusters of granules (small arrows) is evident. In the nucleolar cortex (Cx) very many small homogeneous fibrillar centres (arrows) are present. × 15 000. (Dr F.J. Medina.)
Ex, exine; G, granular component.

where it is thought to play a major role in steroid synthesis. The agranular ER is similarly believed to play a role in glycogen synthesis in liver cells. Experimentally it has been demonstrated with dexamethone (DEX), a potent glucocortoid, that glycogen is deposited in liver cells as discrete $\propto$ and ß particles in close association with agranular ER membranes (Fig 37a, b). The agranular ER forms an extensive system of sacrotubules in skeletal muscle cells, which are related to the binding and releasing of calcium ions for contraction–relaxation processes. It also may participate in the formation of HC1 in the parietal cells of the stomach. Agranular ER cisternae in most cells are frequently found investing lipid droplets (Fig 38).

In a wide variety of cells from vertebrates, invertebrates and plants, morphologically different elaborations are found associated with

47

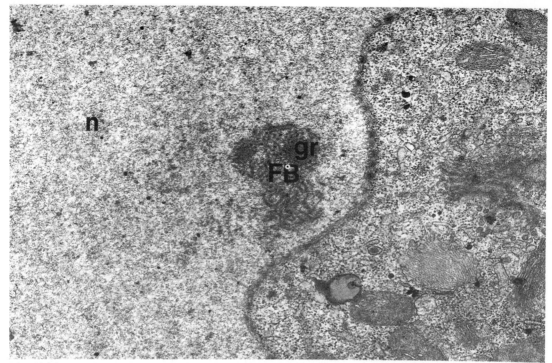

Fig. 30(a) – A nuclear fibrillar body (FB) in the nucleus of the oocyte of sea anemone, *Actinia fragacea* consists of an outer aggregation of fibrils surrounding an inner core of finely granular (gr) material. × 37 000. (Dr A. Larkman.)

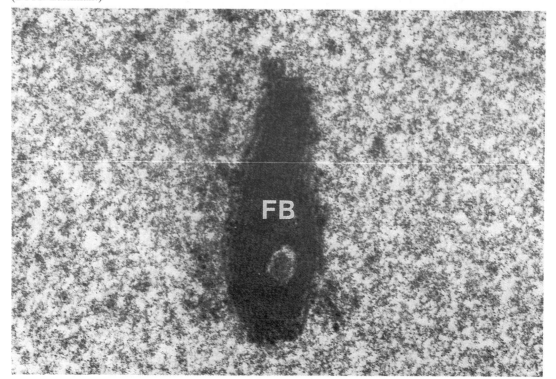

Fig. 30(b) – A nuclear fibrillar body (FB) with a granular core. × 42 000. (Dr A. Larkman.)

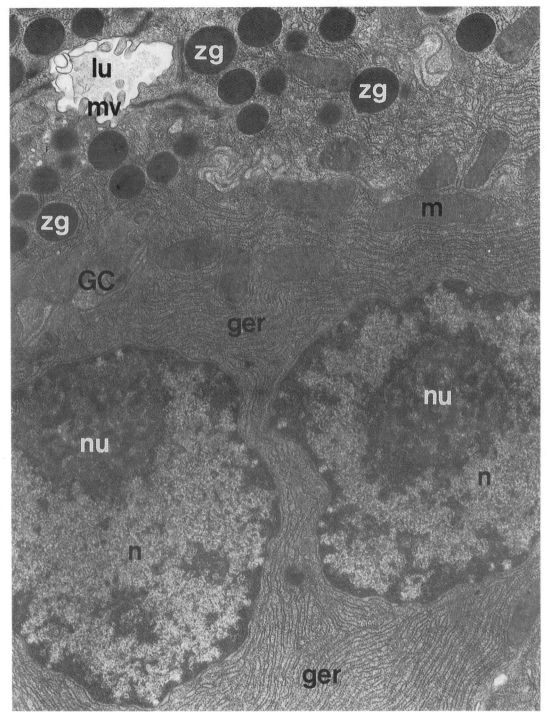

Fig. 31(a) – Acinar epithelial cells of rat pancreas. The luminal surface is provided with a small number of delicate and short microvilli (mv) which protrude into a narrow acinar lumen (lu). The basal two-thirds of the cell is occupied by a dense aggregation of flat cisternae of the granular endoplasmic reticulum (ger) and numerous free ribosomes and polysomes. The Golgi complexes (Gc) are located on the luminal side of the round nucleus (n). Pre-zymogen vacuoles and granules arise in the Golgi complexes. The mature zymogen granules (zg) are spherical and homogeneously electron-dense. × 18 500. (Dr T. Jenkins.)

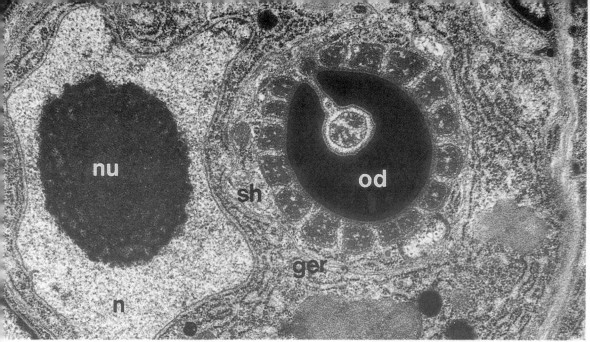

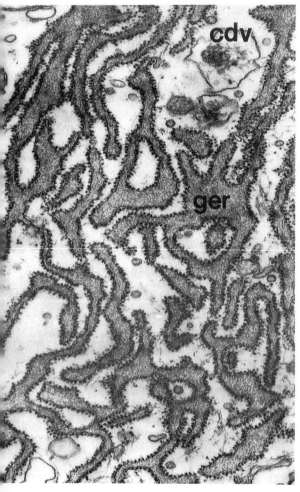

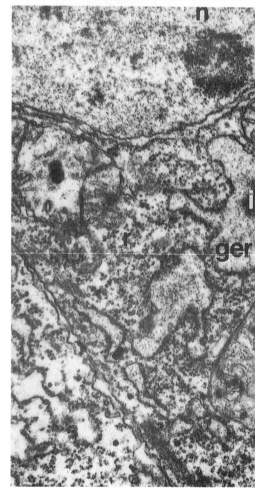

Left

Fig 31(b) – Cross-section through an odontostyle cell of the nematode, *Xiphinema americanum*. The cytoplasm is packed with an extensive granular endoplasmic reticulum (ger) composed of cisternae distended by a flocculent material. × 28 500. (Dr R.F. Carter.)
sh, inner sheath membrane.

Below left

Fig. 32(a) – The granular endoplasmic reticulum (ger) of the collagen gland cell of the bivalve *Mytilus galloprovincialis* is composed of cisternae, which are dilated and contain a microgranular or microfilamentous electron-opaque substance. × 35 000. (Dr L. Vitellaro-Zuccarello, University of Milan.)
cdv, condensing vacuole.

Below centre

Fig. 32(b) – The basal region of the opercular rim cells of the polychaete, *Pileolaria granulata*. The granular endoplasmic reticulum (ger) cisternae are grossly distended and contain intracisternal granules (ig). × 15 000. (Dr A. Bubel.)

Below right

Fig. 32(c) – Cytoplasmic vesicles (v) appear to be derived from the granular endoplasmic reticulum (ger) in the collagen gland cells of *M. galloprovincialis*. These vesicles and granular endoplasmic reticulum transitional elements (arrows) appear to contribute to the formation of condensing vacuoles. × 50 000. (Dr L. Vitellaro-Zuccarello.)

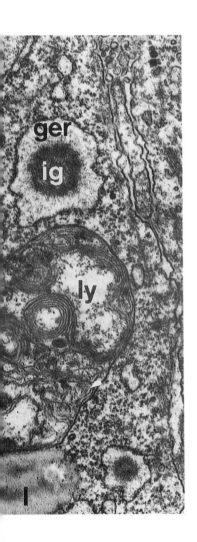

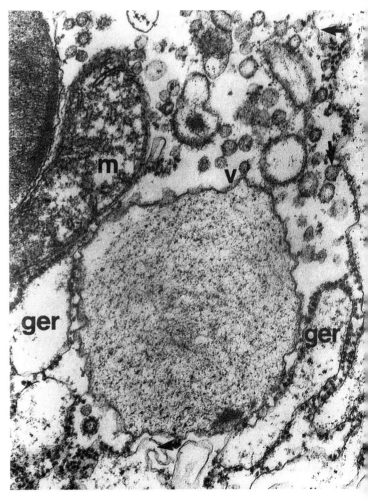

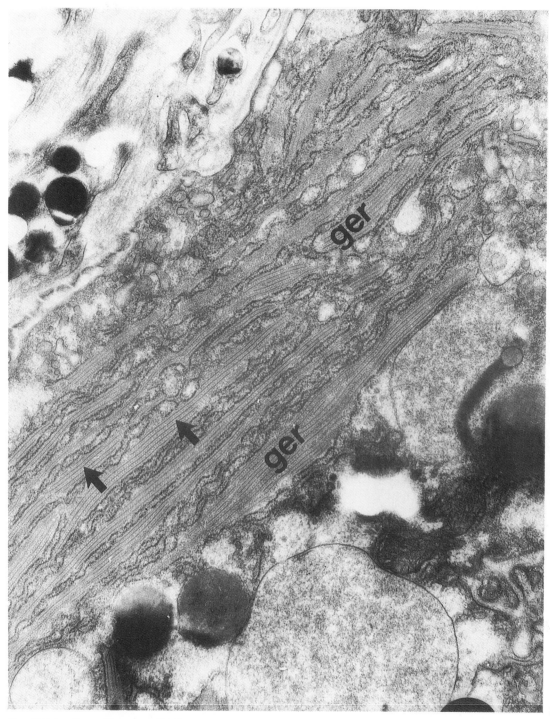

Fig. 33(a) – Crystalline material (haemoglobin molecules) (arrows) in the granular endoplasmic reticulum (ger) of the pore cells of the gastropod, *Biomphalaria glabrata*. × 22 000. (Prof. Dr T. Sminia, Free University.)

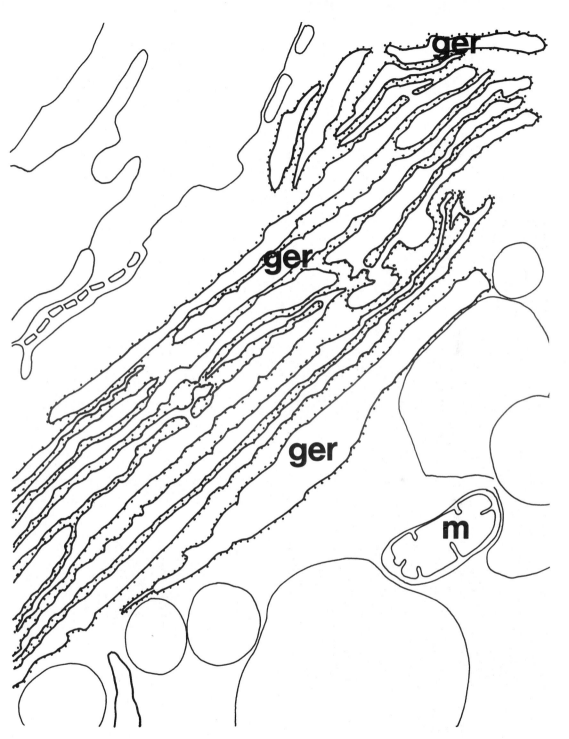

Fig. 33(b) – Diagram of Fig. 33(a).

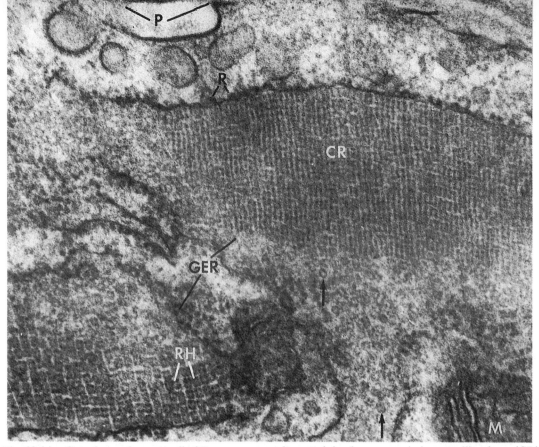

Fig. 33 (c) – Cytoplasm of a pore cell of *Lymnaea stagnalis* containing haemocyanin in crystalline form (CR) in dilated granular endoplasmic reticulum (ger). Haemocyanin molecules are arranged in highly ordered rows (RH). × 58 000. (Prof. Dr T. Sminia.)   P, pores of invaginated plasma membrane.

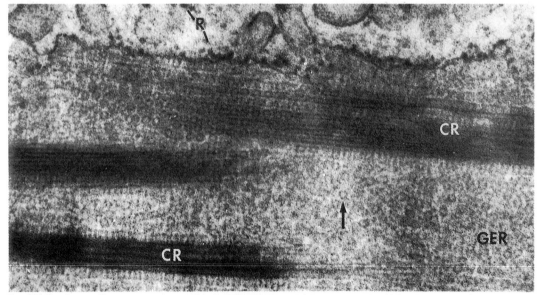

Fig. 33(d) – Pore cell of the gastropod, *Lymnaea stagnalis* with haemocyanin molecules (arrows) in crystalline form (CR) in dilated granular endoplasmic reticulum (ger) cisternae. × 25 000. (Prof. Dr T. Sminia, Free University.)

54

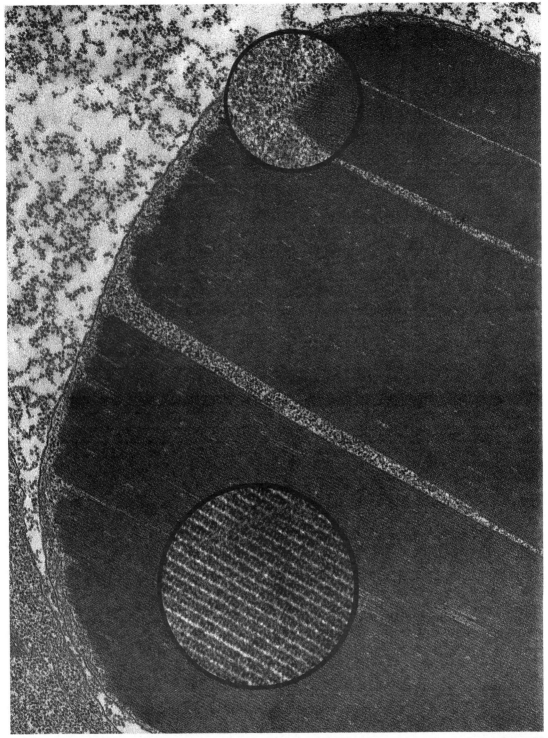

Fig. 34 – A mature blood-cell cyanocyte of the horseshoe crab, *Limulus polyphemus*, packed solidly with haemocyanin crystals. × 45 000. The upper inset shows a polysome attached to the growing end of a crystal. × 80 000. The lower inset reveals the appearance of columns of stacked molecules in the crystal. × 720 000. (Dr W.F. Fahrenbach, Oregon Regional Primate Research Centre.)

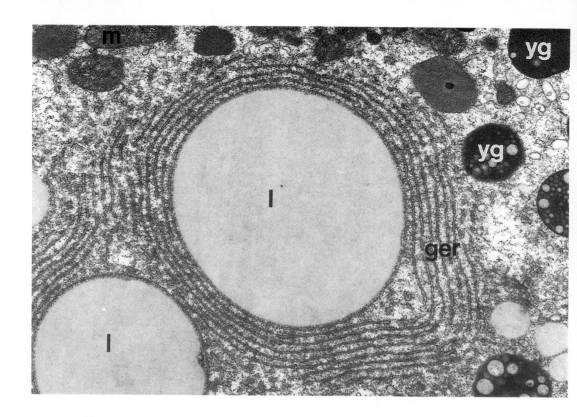

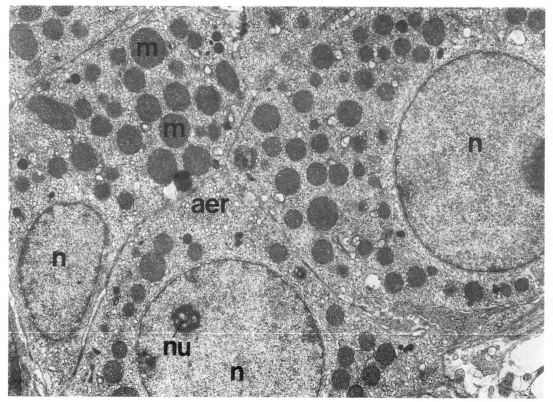

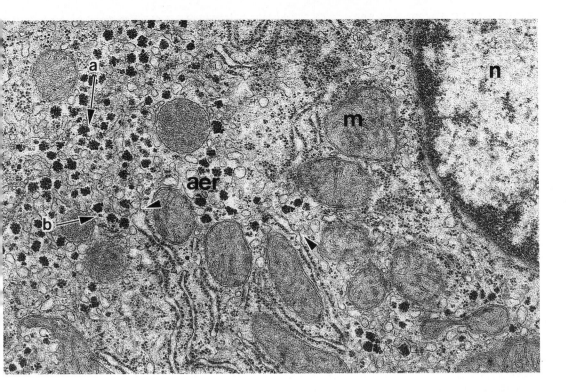

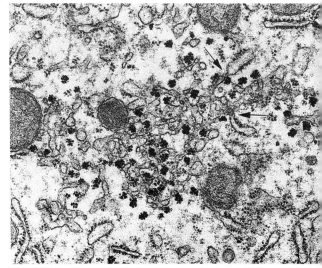

Left above

Fig. 35 – Oocyte of sea anemone, *Actinia fragacea*. A large lipid droplet (l) completely encircled by several cisternae of granular endoplasmic reticulum (ger). × 14 000. (Dr A. Larkman.)
yg, yolk granules.

Left below

Fig. 36 – The cells of the adrenal cortex zone juxta-medullaris of a 12-day-old rat. The nuclei (n) possess a thin rim of heterochromatin at the nuclear periphery. In the cytoplasm, mitochondria (m) possess vesicular cristae and a very dense matrix, and a well-developed agranular endoplasmic reticulum (aer) is present. × 6 600. (Prof. M.M. Magalhaes, University of Porto.)

Above

Fig. 37(a) – The centrilobular hepatocyte cells in rat liver. In the cytoplasm both ∝(a) and ß(b) glycogen particles are evident 2 hours after dexamethasone injection. The particles are dispersed throughout the cytoplasm in regions rich in agranular endoplasmic reticulum, whereas other organelles are excluded from the agranular endoplasmic reticulum regions. In the glycogen-rich areas there is continuity of granular endoplasmic reticulum with agranular endoplasmic reticulum (arrow heads). × 28 000. (Dr W.G. Jerome, Bowman Gray School of Medicine.)

Fig. 37(b) – Four hours after dexamethasone administration, the characteristic spatial association of agranular endoplasmic reticulum and glycogen is quite evident in the centrilobular hepatocytes, as is the continuity between granular and agranular endoplasmic reticulum (arrows) in the glycogen-containing regions of the cytoplasm. The agranular endoplasmic reticulum, as with periportal hepatocytes, becomes more tubular in appearance after dexamethosone treatment. × 36 500. (Dr. W.G. Jerome.)

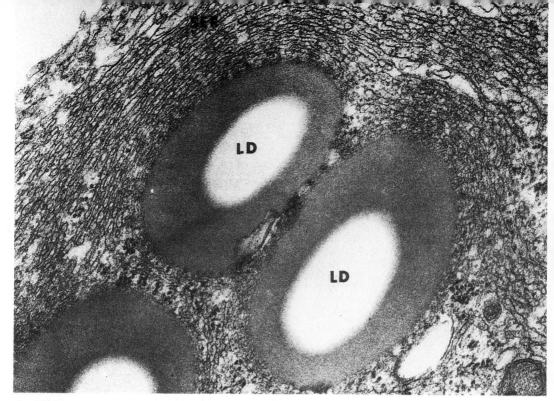

the ER. In vertebrates, 'tubular bodies' 'tubu-loreticular structures', and 'undulating tubules', are found, whereas 'paracrystalline lattices', 'paracrystals of tubular ER', 'sub-microvillar cisternae', 'crystalline lattices' and 'undulating ER' (Fig 39a–d) are observed in invertebrates, and 'lattice bodies', 'ER aggregates' and 'lattice-like membrane structures' are reported in plants. Such structures are thought to be responsible for the synthesis and storage of cholesterol, metabolism/transport of photopigment in transducing membranes, or storage of vitamin A compounds, or to be a means of storing a large agranular ER surface in a small space. In the gastropod, *Spirilla neapolitana*, an array of undulating ER cisternae appear in sperm as a normal development of spermatozoa. The undulating cisternae are continuous with the membranes of both granular and agranular ER (Fig 39).

In most instances, the morphological variations associated with the ER are believed to be manifestations of pathological conditions. In the toad *Bufo arenarum*, microtubular formations are found in the granular ER of cells of the pars intermedia, following extirpation of the pars distalis (Fig 40a, b). Similar intra-cisternal microtubular formations are also found in differentiating plant cells.

Above
Fig. 38 – The ovarian granulosa cells of the rat containing in the cytoplasm lipid droplets (LD) encircled by cisternae of agranular endoplasmic reticulum (SER). × 60 000. (Dr E. Anderson, Harvard Medical School.)

Right above
Fig. 39(a) – An array of undulating endoplasmic reticulum in close association with a Golgi complex (Gc) in the spermatid of the mollusc *Spurilla neapolitane*. There is continuity between the cisternae of undulating and normal endoplasmic reticulum (arrows). × 24 000. (Dr K.J. Eckelbarger, Harbor Branch Foundation.)
A, axoneme.

Right below
Fig. 39(b) – Diagram of Fig. 39(a).

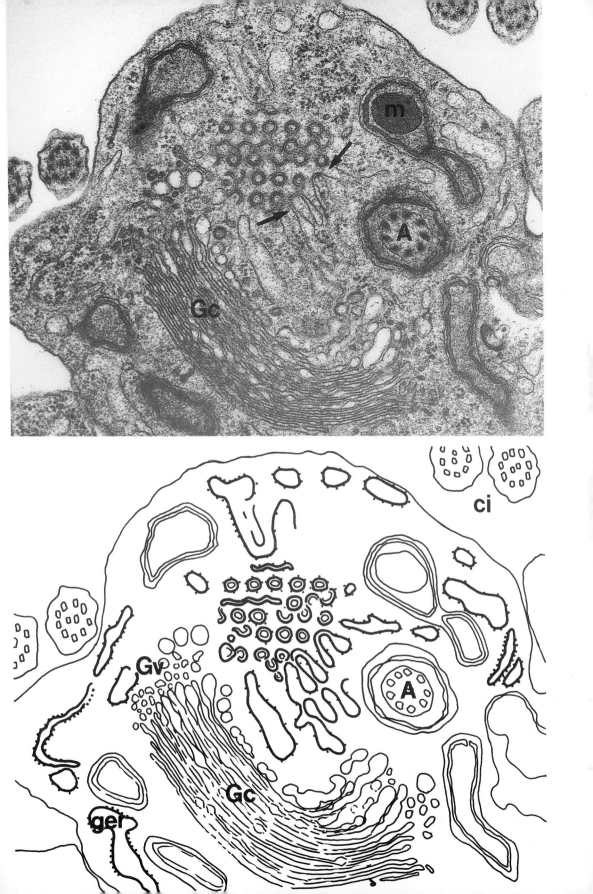

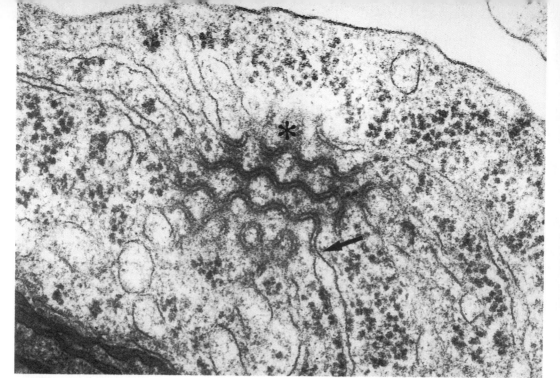

Fig. 39(c) – A single cisternae of undulating endoplasmic reticulum continuous with normal endoplasmic reticulum (arrows). An electron-dense flocculum is associated with undulating endoplasmic reticulum (*). × 52 500. (Dr K.J. Eckelbarger.)

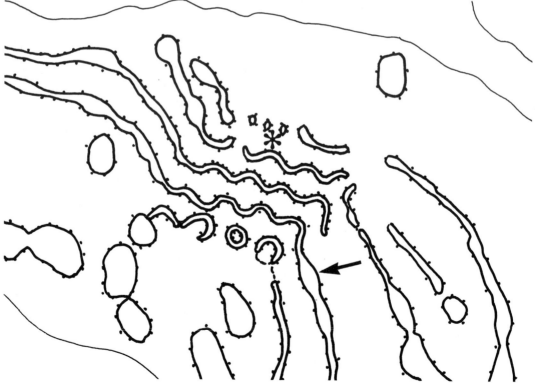

Fig. 39(d) – Diagram of Fig. 39(c).

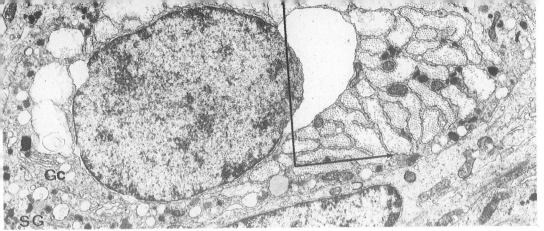

Fig. 40(a) – Four weeks after the extirpation of the pars distalis from the toad, *Bufo arenarum*, the granular endoplasmic reticulum of the pars intermedia possesses dilated cisternae packed with microtubules. × 9 500. (Dr E.M. Rodriguez, University of Chile.)   Gc, Golgi complex; sg, secretory granules.

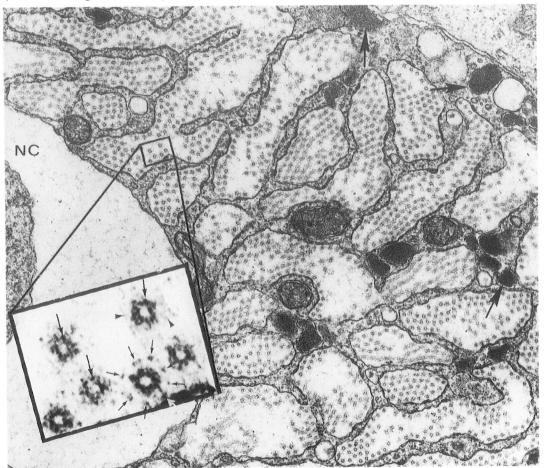

Fig. 40(b) – The microtubular structures occupy the lumen of dilated granular endoplasmic reticulum cisternae, but are absent from the dilated region of nuclear cisternae (NC). In the cytoplasm, secretory granules (arrows) are present. × 35 500.
Inset: In cross-section the wall of the microtubules appears to be formed of small particles (large arrows). Satellite particles (small arrows) are arranged in an orderly manner around a microtubule. From the surface of microtubules a thread-like material projects (arrow heads). × 265 000. (Dr E.M. Rodriguez.)

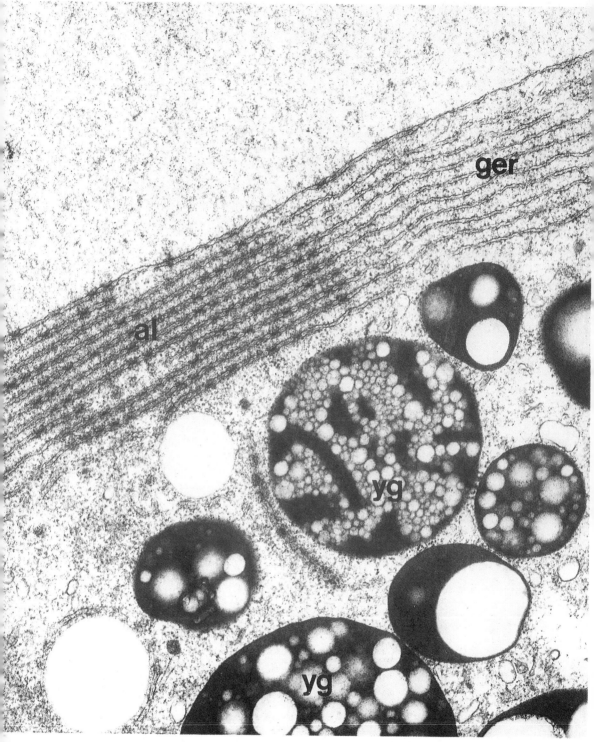

Fig. 41(a) – A stack of annulate lamellae (al) among a perinuclear band of endoplasmic reticulum in a vitellogenic oocyte of the sea anemone, *Actinia fragacea*. The membranes of the annulate lamellae are continuous with those of the endoplasmic reticulum. × 59 000. (Dr A. Larkman.)

ANNULATE LAMELLAE

These consist of from three to ten parallel smooth membranes. Each pair of membranes contains a variable number of regularly spaced pores. Annulate lamellae are considered to represent a specialized form of endoplasmic reticulum. They are found in a number of cell types, most commonly in developing male and female germ cells, certain embryonic cells and in tumour cells. In vitellogenic oocytes, annulate lamellae are always found associated with the ER, with large accumulations found among perinuclear bands of ER. In such a situation, the membranes of the two are continuous (Fig 41). Each annulate lamella closely resembles a portion of the nuclear envelope, its numerous pores are similar to nuclear pores (Fig 41). The membranes of annulate lamellae extend for a short distance beyond the pore complexes and may be studded with ribosomes. Ribosomes are also found between individual lamellae. Despite the widespread occurrence of annulate lamellae, there is no general agreement as to their mode of formation or function. Hypothetically, they could transfer genetic material from the nucleus to the cytoplasm.

GOLGI COMPLEX

The Golgi complex (synonyms: Golgi zone; region; body; apparatus; substance; dictyosome) is a heterogenous organelle present in most eukaryotic cells. It is especially prominent in cells secreting either proteins or complex polysaccharides. The location, size and development of these organelles vary from cell to cell and also with the physiological state of the cell. In cells that have a polarized structure, such as thyroid cells, exocrine pancreatic or mucous cells, the Golgi complex is located between the nucleus and the apical cell surface (Fig 42a, b). However, in large neurones, such as spiny ganglion cells, some invertebrate tissues and plant cells, individual organelles are found scattered throughout the cytoplasm, and their distribution does not seem to be ordered or localized in any particular manner. The number per cell can vary from about 50 per cell in liver cells to several hundred, as in the tissues of the corn root, to a single organelle in some algae. The number of Golgi complexes is either maintained at a constant level, or even increased as cells grow and divide. In a number of tissues, the total number of Golgi complexes per cell has been shown to increase several-fold as the physiological state of the tissue changes.

The form of Golgi complexes varies greatly in different cells. Their shape may differ within a cell as a result of varying metabolic activity and the developmental state of the cell. It may range from pleomorphic masses which lack symmetrical properties to hollow cones, cylinders and bowls (Figs 43, 44a–f). The variations in shape do not, however, appear to influence the polar arrangement of the stacked

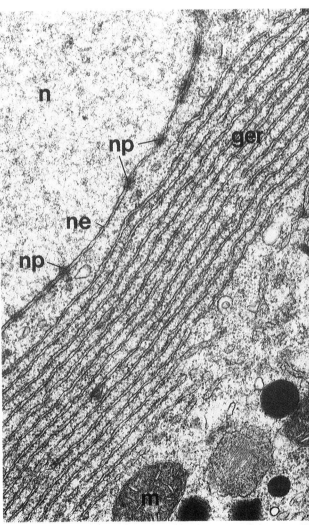

Fig. 41(b) – A part of a perinuclear mass of endoplasmic reticulum (ger) in a vitelline oocyte of *A. fragacea* which occur close to the nuclear envelope (ne). × 18 500. (Dr A. Larkman.)

63

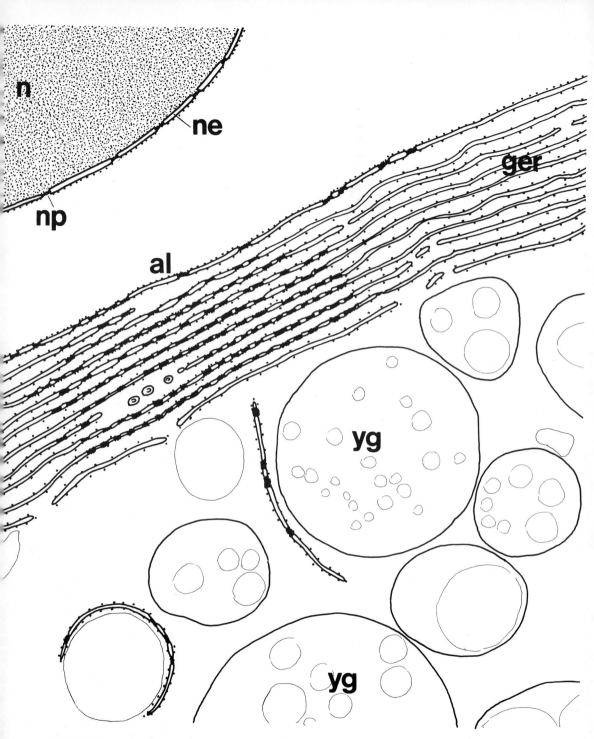

Fig. 41(c) – Compilation diagram of Figs. 41(a,b).

cisternae. Each organelle is, however, composed of a stack of parallel plate-like cisternae (also called saccules or lamellae), which, in section, appear as flattened sacs bounded by the usual three-layered unit membrane. The number of cisternae in a stack varies, stacks of three to seven are common in higher animal and plant cell types, with between 10 and 20 in some unicellular algae. Bundles of fibrils can often be seen to extend across the inter-cisternal space and may serve as strengthening elements maintaining the regular spacing between individual cisternae. The cisternae are usually dilated towards their periphery, and there appear to be vesicles, either attached to or situated close to their marginal regions, as if they had been budded off from the rim (Figs 42, 43). The cisternae of an individual Golgi complex are not all equivalent they usually show a definite polarization. There is a prox-

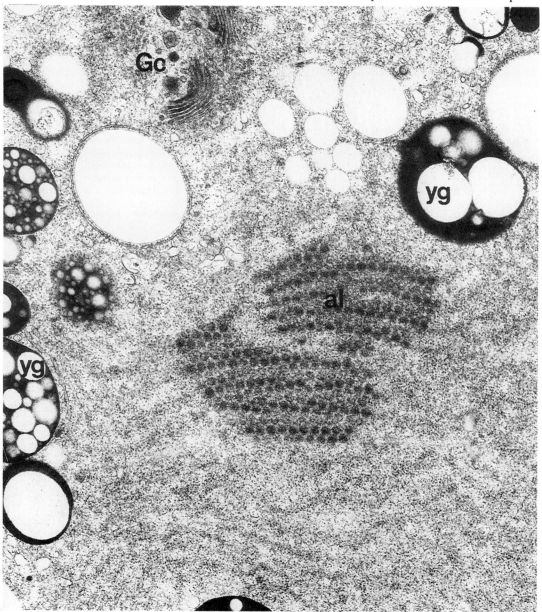

Fig. 41(d) – A small area of annulate lamellae (al) located in a mass of endoplasmic reticulum in the peripheral cytoplasm of the oocyte of *A. fragacea*. × 26 500. (Dr A. Larkman.)

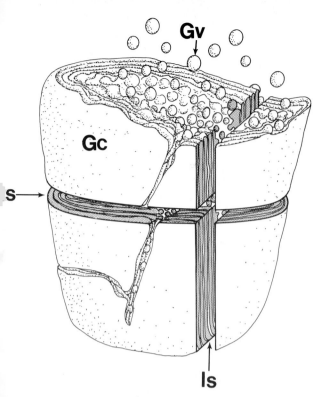

Left

Fig. 42(a) – A diagrammatic representation of the three-dimensional, transverse sectional (ts) and longitudinal sectional (ls) structure of a Golgi complex (Gc). In some cells the Golgi complex forms a cup-like structure that is made up of several elements, each consisting of a stack of eight to ten cisternae. The organelle looks like a cracked cup with secretory vesicles (Gv) budding off from the lip of the cup.

Below

Fig. 42(b) – Transverse section through a sea urchin, *Echinus sp.*, oocyte, exhibiting a circular Golgi-complex profile. A structural polarity is evident in the organelle with the forming face (FF) cisternae closely associated with vesicular elements and cisternae of an extensive endoplasmic reticulum (ger), whereas numerous secretory vesicles (gv), interpreted as intermediate stages in the formation of secretory granules (sg), are associated with the maturing face (FM). × 22 000. (Dr G. Charles.)

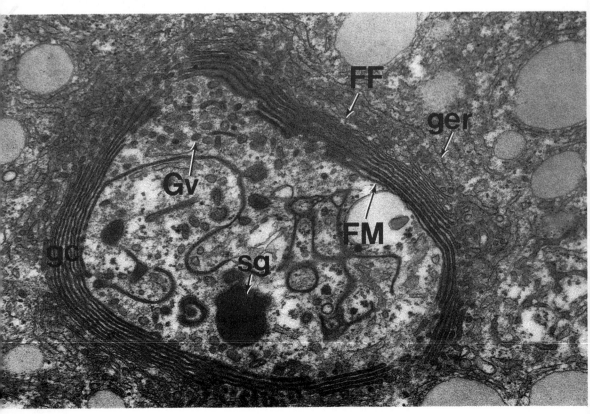

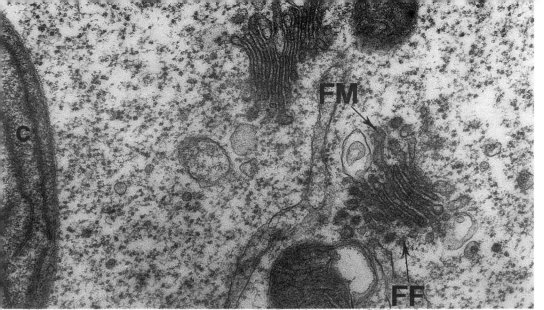

Fig. 43(a) – An attached zoospore of the alga *Polysiphonia sp.* in which the secretion of an adhesive substance is completed. In such a cell, there is little evidence of a structural polarization of cisternae within the Golgi complex. The lumina of both forming (FF) and maturing face (FM) cisternae are of similar dimension. Distended cisternae and secretory vesicles are either lacking or are few in number. × 30 000. (Dr R.L. Fletcher, Portsmouth Polytechnic.)

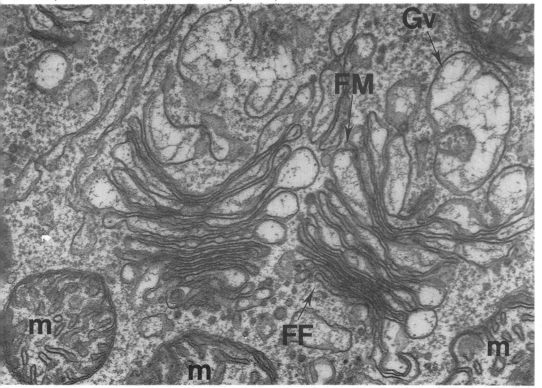

Fig. 43(b) – During the process of *Polysiphonia* zoospore attachment, there is a clearer indication of polarity in the cisternae of a Golgi complex. The organelle responds to the increased demand for secretory product, i.e. adhesive, by marked changes in its organization. There is a conspicuous enlargement of the maturing face (FM) cisternae and distension of their marginal tips, from which region secretory vesicles (GV) containing adhesive material are pinched off. × 34 000. (Dr R.L. Fletcher.)

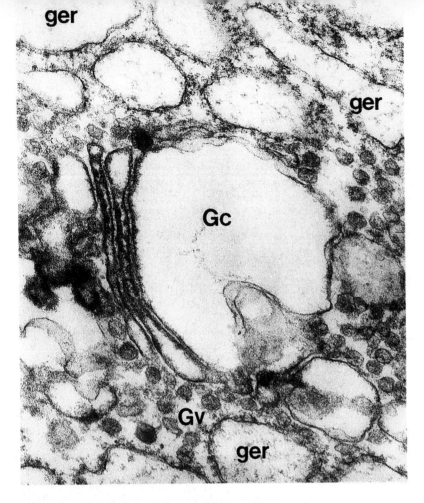

imal or forming face which is close to the nuclear envelope or to the endoplasmic reticulum, and a distal or maturing face associated with the formation of secretory vesicles. The forming face usually has a convex profile in thin sections, whilst the maturing face has a distinctly concave profile, though the degree of curvature varies with the plane of section and with the cell type.

The Golgi complex is an extremely dynamic organelle performing many key functions in the cell. A general feature of the organelle in both animal and plant cells is the formation and packaging of polysaccharides, proteins, glycoproteins and glycolipids for export from the cells. The pattern of secretion of a particular extracellular product seems to be similar in both animal and plant cells. For example, during the secretion of glycoproteins, secretory proteins synthesized on ribosomes of the granular ER are transferred into the cisternae of the ER, and then into smooth vesicular

elements which are carried to the forming face of the Golgi complex. In the Golgi cisternae the protein is glycosylated, i.e. polysaccharide is added to complete the glycoprotein. At the maturing face, Golgi cisternae are more swollen and it is from here that the majority of secretory vesicles containing glycoprotein are shed into the cytoplasm. In protein-secreting cells, secretory proteins are concentrated and packaged by the Golgi complex to be released as small Golgi vesicles, which may coalesce into larger vacuoles or condensing vacuoles. Golgi vesicles/vacuoles in cells give rise to protein zymogen granules, which are discharged from cells by reverse pinocytosis. During the latter process the membrane of the Golgi vesicle/vacuole fuses and becomes continuous with the plasma membrane. Simultaneously, the contents of the vesicle/vacuole are released. In contrast, in goblet cells, membrane-bound glycoprotein mucigen granules pass through gaps in the plasma

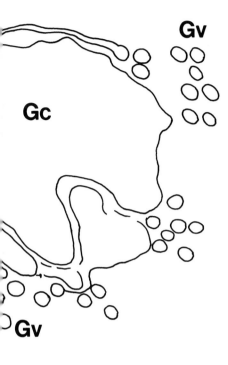

Gv

Gc

Gv

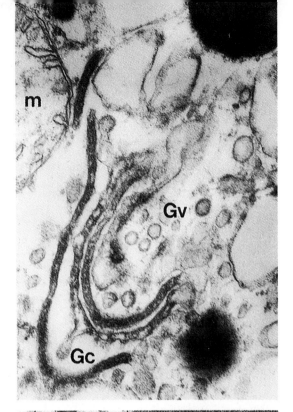

m

Gv

Gc

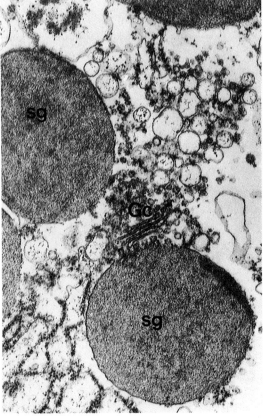

sg

Gc

sg

Above left
Fig. 44(a) – The Golgi region of an enzyme gland cell in the foot of the bivalve, *Mytilus galloprovincialis*, incubated in diethyldithiodicarbamide (DDC) before exposure to DOPA. × 65 000. (Dr L. Vitellaro-Zuccarello, University of Milan.)

Above centre
Fig. 44(b) – Diagram of Fig. 44(a).

Above right
Fig. 44(c) – The Golgi region of an enzyme gland cell in the foot of *M. galloprovincialis* incubated for phenol oxidase activity. × 50 000. (Dr L. Vitellaro-Zuccarello.)

Right
Fig. 44(d) – The close relationship of a Golgi complex and a secretory granule (sg) in a collagen gland cell in the foot of *M. galloprovincialis*. × 31 000. (Dr L. Vitellaro-Zuccarello.)

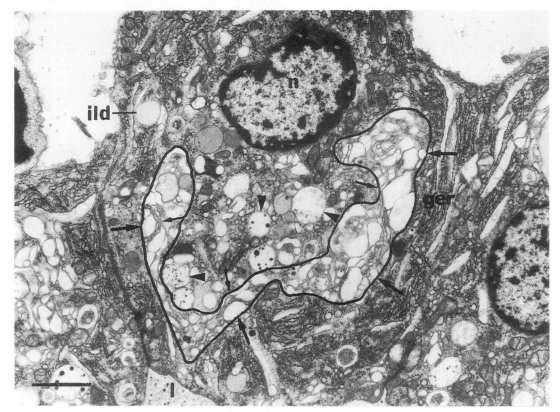

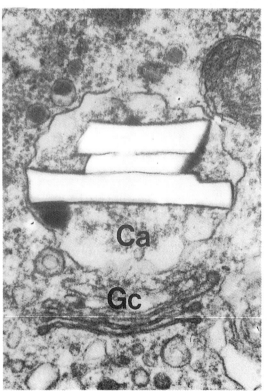

Above
Fig. 44(e) – A U-shaped Golgi apparatus in the milk-secreting epithelial cells of bovine mammary gland. The secretory vesicles (arrow heads) predominate on the maturing face (trans) (small arrows) of the Golgi complex, and the granular endoplasmic reticulum (ger) predominates on the forming face (large arrows). × 8 000. (Dr D.P. Dylewski, Virginia State University.)

Left
Fig. 44(f) – A calcifying vesicle (Ca) in ephyra rhopalium of jellyfish, *Aurelia aurita*. The statolith (calcium sulphate) 'ghosts', are closely associated with a Golgi complex, and small membrane-bound dense-core vesicles. × 225 000. (Dr D.B. Spangenburg, University of Colorado.)

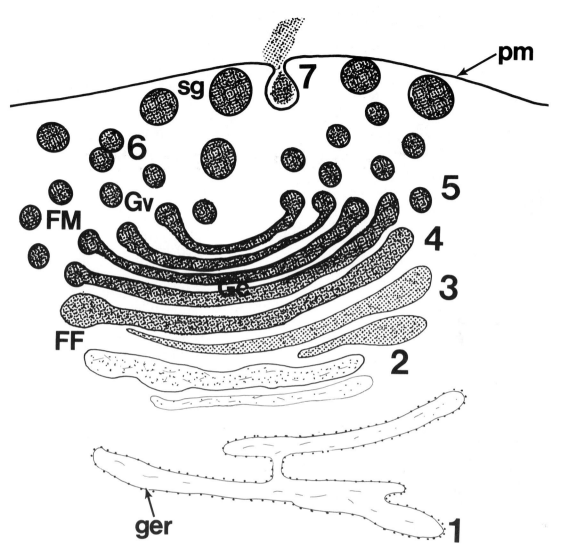

Fig. 45 – Diagrammatic representation of the dynamic nature of the Golgi complex. (1) Secretory protein is synthesized on the ribosomes of the granular endoplasmic reticulum (ger) and the nascent protein is transferred into the cisternae of the granular endoplasmic reticulum. (2) The synthesized protein is transferred to the forming face (FF) cisternae by means of small vesicles which are blebbed off the granular endoplasmic reticulum. At the forming face, new cisternae are added by the input of new membrane. The bounding membrane of the forming face cisternae is approximately 6 nm thick and resembles that of the endoplasmic reticulum. (3) Protein material within the cisternae becomes concentrated during its passage from the forming to the maturing face (FM) of the organelle. Also, the addition of saccharide residues to the protein (glycosylation) may occur to give glycoproteins. (4) In the Golgi complex (Gc) the cisternae mature, undergo differentiation and move progressively across the stack towards the maturing face. At the maturing face the bounding membrane of the cisternae is 8 nm thick and resembles the plasma membrane (pm). (5) At the maturing face, dilated cisternal tips contain concentrated secretory product, e.g. protein, glycoprotein, are nipped off to form secretory vesicles (Gv) bounded by an 8 nm thick membrane. (6) Secretory vesicles may fuse with one another to give rise to larger secretory granules. (7) Fusion of the secretory granule membrane with the plasma membrane results in the incorporation of new material into the plasma membrane and in the release of the secretory product.

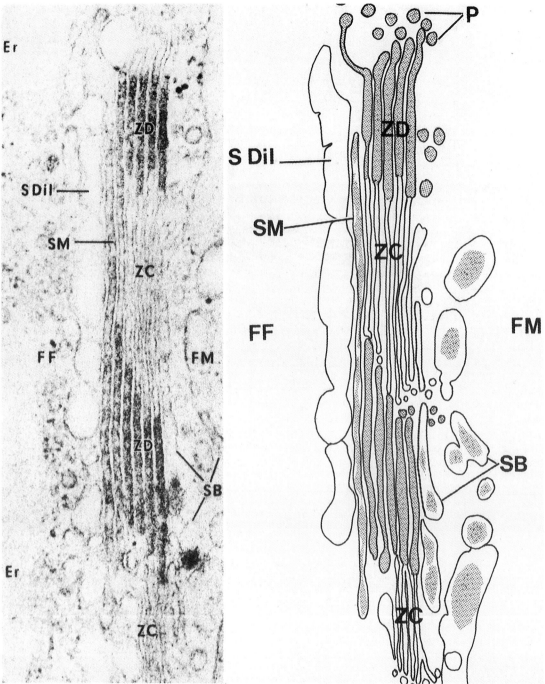

Above left

Fig. 46(a) – The Golgi complex of a multifid gland secretory cell of the snail, *Helix sp*. The periodic acid–thiocarbohydrazide–silver proteinate-staining reaction and proteolytic enzyme digestion techniques reveal that within a single organelle there is a functional differentiation of cisternae into clear (ZC) and dense (ZD) zones. These correspond to the simulta-

neous formation, at the maturing face (FM), of two secretory products, namely, glycoproteins (SB) and proteins (P). × 30 000. (Dr L. Ovtracht, University of Paris.)

Er, endoplasmic reticulum; s Dil, distended forming face cisternae; SM, mixed cisterna.

Above right

Fig. 46(b) – Diagram of Fig. 46(a).

membrane to the outside of the cell, where they rupture allowing mucus to escape. In hypersecretory cells, Golgi complexes possess swollen cisternae that produce numerous vesicles, which ultimately release their contents from the cell. There is, in addition, frequently a temporary accumulation of Golgi secretory granules between the cytoplasm and the plasma membrane. In inactive secretory cells, Golgi complexes are inconspicuous, with few Golgi secretory vesicles/vacuoles in the cytoplasm (Fig 43a, b).

The evidence that links Golgi complexes with the secretion of a particular extracellular product comes from histochemical and autoradiographic studios. However, the interrelationship of secretory vesicles to the Golgi complex, and their direction of movement towards the cell plasma membrane, can only be inferred indirectly. There is the remote possibility that vesicles may traverse the cell from the plasma membrane to the Golgi complex.

Chemically, Golgi cisternal membranes are intermediate between that of the ER and plasma membrane. Their transitional nature is also evident from morphological studies. The membranes of the forming-face cisternae, which border on the membranous components of either the nucleus or the ER, are on average 6 nm thick, whereas those of maturing-face cisternae are, on average, 8 nm thick, and similar to the plasma membrane. Secretory vesicles leaving the maturing-face cisternae are bounded by an 8 nm-thick smooth trilaminar membrane. At the forming-face, new cisternae are believed to be added continuously, by an input of membrane material from either the nuclear envelope or the ER. In the organelle, as the cisternae mature they move progressively across the stack and undergo differentiation to become more similar in structure and composition to the plasma membrane. It has been estimated that in mucus-producing mammalian intestine goblet cells, it takes 40 minutes for a stack of cisternae to be completely turned over and transformed into secretory vesicles. This implies that Golgi cisternal membranes are very rapidly transformed into plasma membranes. The process of turnover is, however, likely to occur more slowly in cells which are synthetically less active. The transfer of membrane-bound vesicles from the Golgi complex to the plasma membrane is responsible in some cells for the addition of surface plasma membrane during growth, i.e. pancreatic exocrine cells (Fig 45).

In some cells, different cisternae in the same organelle produce distinct kinds of secretory product; for example, in wheat leaf stomatal cells, some cisternae give rise to vacuolar fluid, whereas others simultaneously are involved in the secretion of materials destined for the cell wall. In addition, in the multifid gland cells of the snail, *Helix,* glycoproteins are secreted

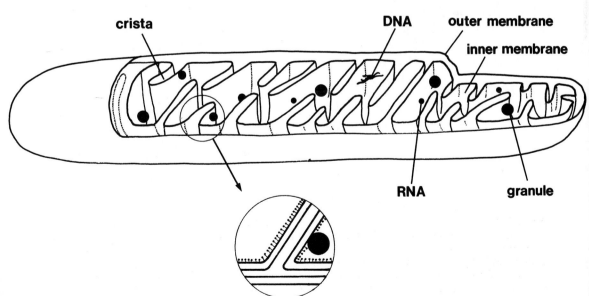

Fig 47(a) – Three-dimensional diagram of a mitochondrion.

independently of proteins into the cytoplasm (Fig 46). The simultaneous formation of more than one type of product by a single Golgi complex indicates that there must be some degree of compartmentalization within the organelle.

In addition to its anabolic functions, the Golgi complex plays a part in catabolic activities in cells, particularly in the formation of lysosomes. Primary lysosomes, containing hydrolytic enzymes, may arise either from the Golgi complexes, according to the route proposed for secretory proteins, or directly from the ER. When primary lysosomes containing hydrolytic enzymes fuse with pinocytotic vesicles or autophagic vacuoles in a cell, they give rise to secondary lysosomes or digestive vacuoles.

In cells, Golgi complex secretions are not always restricted to organic materials. The organelle may, in some plant and animal cells, concentrate silicon, calcium, and water. There is also the close association between the Golgi complex and calcium sulphate-containing vacuoles in some invertebrates (Fig 44f).

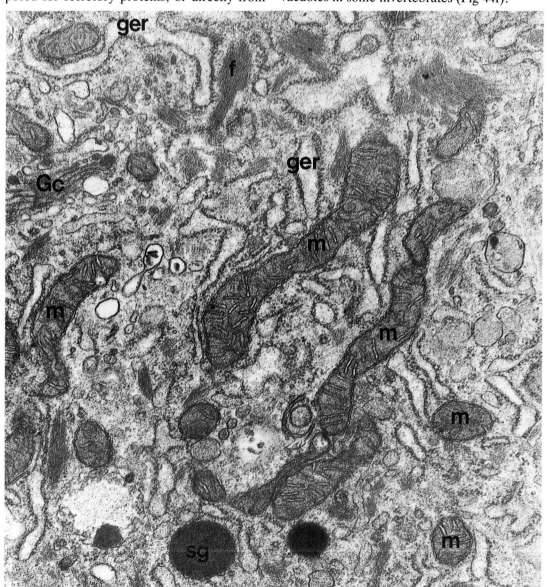

Fig. 47(b) – Typical mitochondria (m) with transverse cristae in the opercular plate cells of the polychaete *Pileolaria granulata*. × 22 000. (Dr A. Bubel.)

74

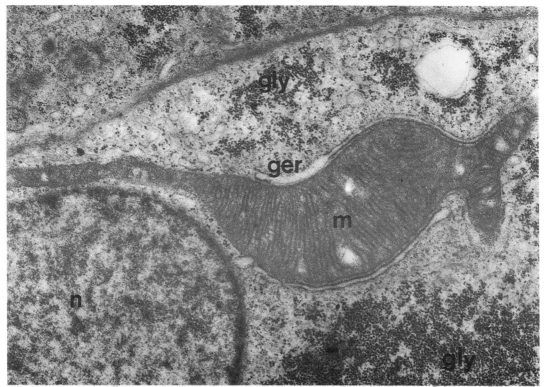

Fig. 48(a) – A giant mitochondrion (m) with a branched profile in the base of a human endometrial glandular cell. × 18 500. (Dr T. Coaker, Royal Victoria Infirmary, Newcastle.)

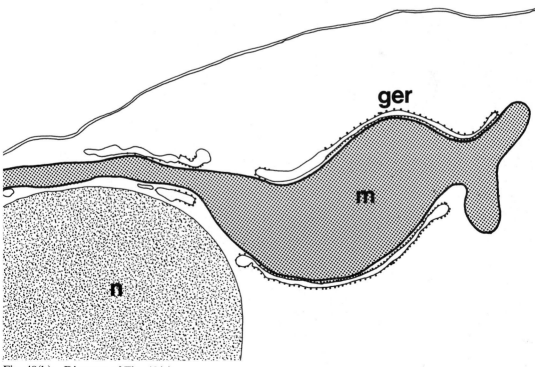

Fig. 48(b) – Diagram of Fig. 48(a).

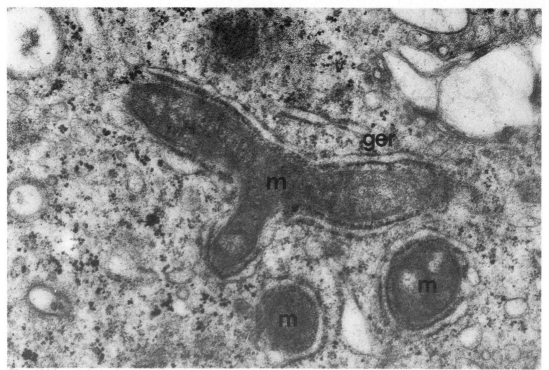

Fig. 48(c) – A branched profile of part of a giant mitochondrion (m) invested by granular endoplasmic reticulum. × 36 000 (Dr T. Coaker.)

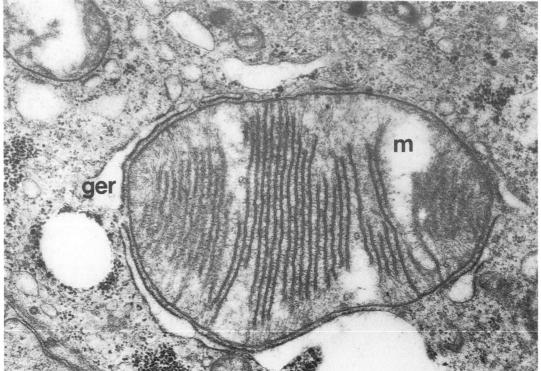

Fig. 48(d) – A giant mitochondrion (m) with closely packed cristae and a close investment of granular endoplasmic reticulum in the base of a human endometrial glandular cell. × 24 000. (Dr T. Coaker.)

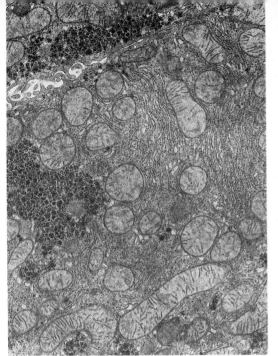

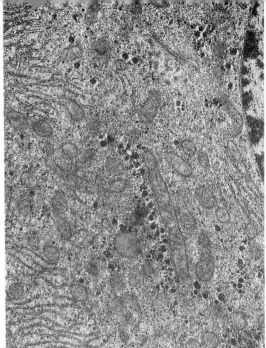

Fig. 49(a) – A periportal cell from mouse liver containing large to medium mostly round mitochondria. × 10 500. (Dr S. Kanamura, Kansai Medical University.)

Fig. 49(b) – A perihepatic cell from mouse liver containing small, round to slender mitochondria. × 10 500 (Dr S. Kanamura.)

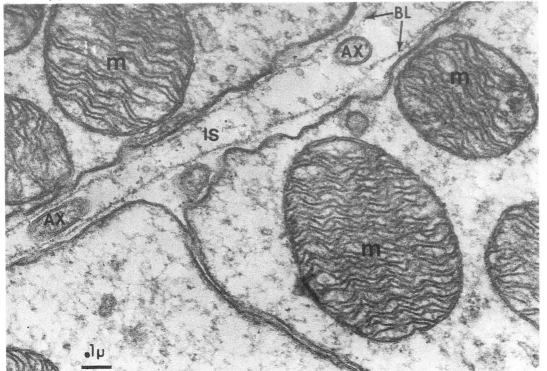

Fig. 50 – A portion of the secretory cells of the accessory boring gland of the gastropod, *Urosalpinx sp*. The cells contain mitochondria (m) with wavy cristae, which are nearly parallel to each other. × 73 500. (Dr M.U. Nylen, National Institute of Dental Research, Bethesda.)

77

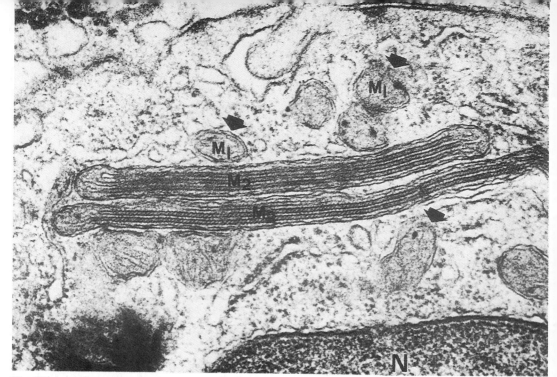

Fig. 51(a) – A fragment of a pneumocyte of the newt *Triturus alpestris,* containing two atypical mitochondria (M₂) with obliquely septate cristae, and normal mitochondria (M₁) (arrow). × 59 000. (Dr W. Witalinski, Jagiellonian University.)

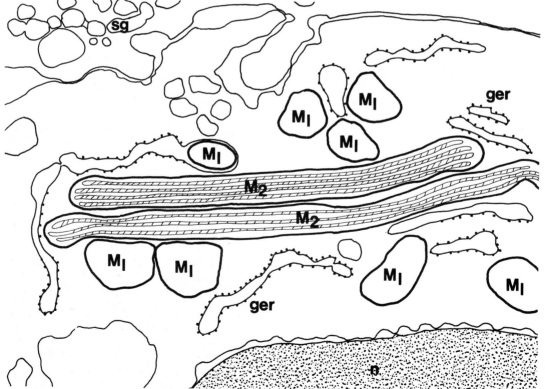

Fig. 51(b) – Diagram of Fig. 51(a).

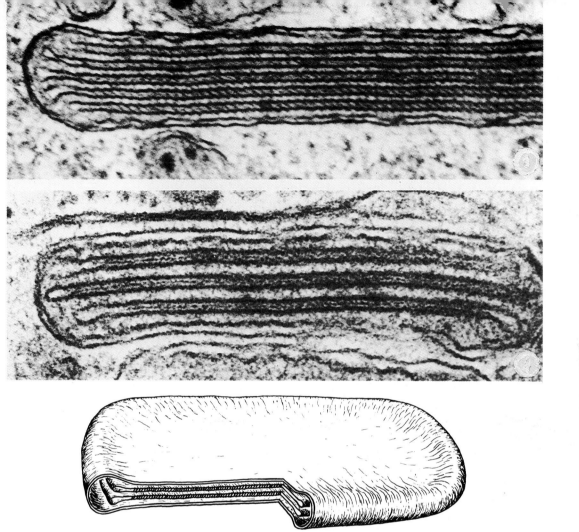

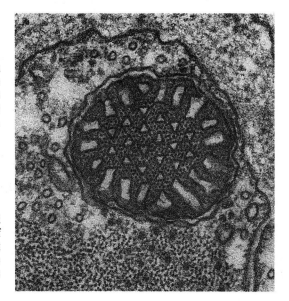

Above top

Fig. 51(c) – A portion of a mitochrondrion from a pneumocyte of *T. alpestris* with five obliquely septate cristae. A distinct oblique striation is evident in the two lower cristae, but the swollen margins of the cristae are devoid of striations. × 124 000. (Dr. W. Witalinski.)

Above centre

Fig. 51(d) – Diagramof Fig. 51(c).

Right

Fig. 52 – A mitochondrion with centrally located triangular cristae, found in the abdominal muscle of the crayfish, *Procambarus clarkii*. In the matrix between the cristae, transversely sectioned fibrils are evident. × 79 000. (Dr T. Komuro, Ehime University.)

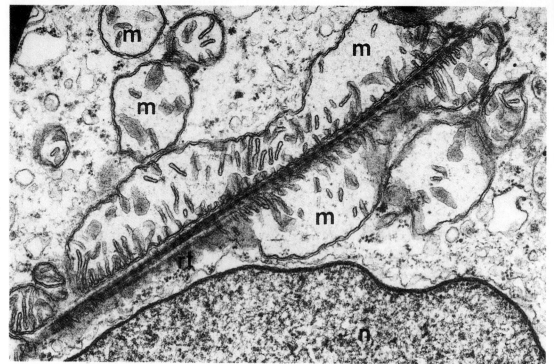

Fig. 53(a) – A light-sensitive sensory cell of the cerebral ocellus of the polychaete *Spirobranchus giganteus*, with mitochondria (m) in close association with a rootlet (rt) (ciliary?). The cristae appear to be aligned with the rootlet bands. × 45 000. (Dr R.S. Smith, James Cook University.)

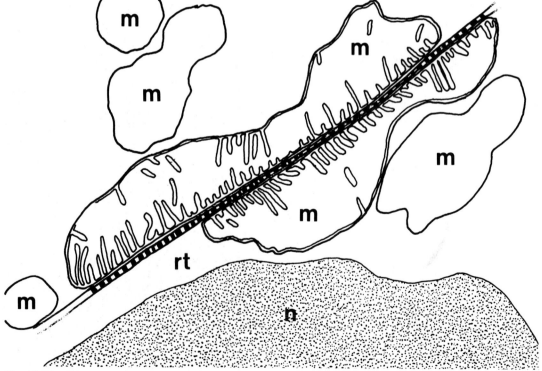

Fig. 53(b) – Diagram of Fig 53(a).

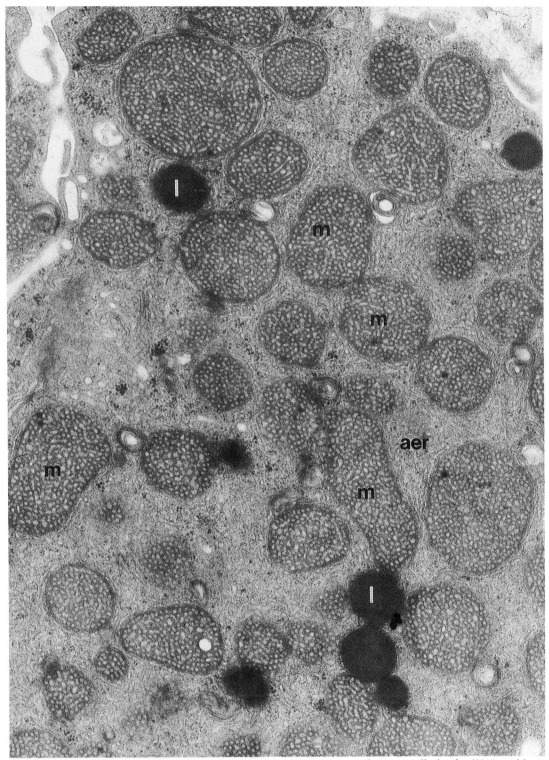

Fig. 54(a) – Steroid-hormone-secreting cells of the adrenal cortex zona juxtamedullaris of a 4½-day-old rat, containing mitochondria (m) with vesicular/tubular cristae, and predominantly agranular endoplasmic reticulumin (aer) in the cytoplasm. × 21 500. (Prof. M.M. Magalhaes, University of Porto.)

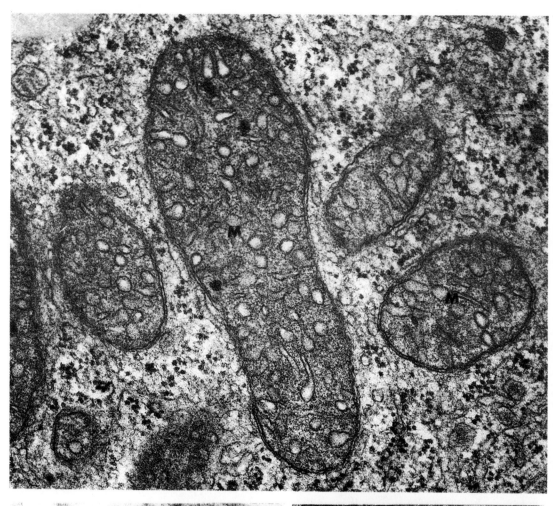

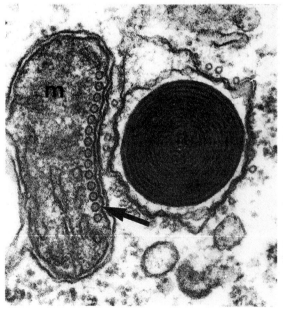

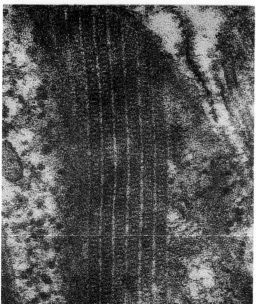

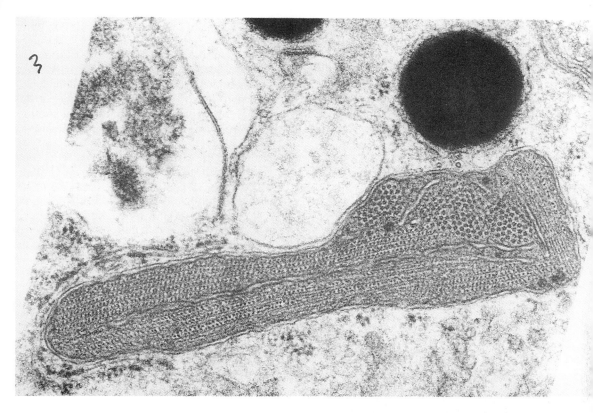

2

MITOCHONDRIA

Mitochondria are organelles that are the major providers of energy in both animal and plant cells. They contain a large number of enzymes and coenzymes that enable them to recover the energy contained in food material and convert it by phosphorylation into the high-energy phosphate bond of adenosine triphosphate (ATP). They are found in all aerobically respiring cells with the exception of bacteria. Mitochondria possess a specific structural organization. Each organelle is bounded by a smooth outer membrane about 6 nm thick, which is separated from an inner membrane component about 6 nm thick by an electron-lucent space 8 nm thick. The inner membranous component is composed of a series of infoldings called cristae that project into the central chamber, which is usually filled by a homogeneous or finely granular matrix (Fig 47a,b). The matrix contains the enzymes of the tricarboxylic acid cycle (except for succinate dehydrogenase), enzymes of ß-oxidation, the enzymes carbamoyl phosphate synthetase and ornithine carbamoyl transferase, which relate to urea formation, and others.

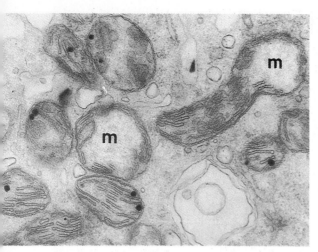

Fig. 57(a) – Mitochondria (m) with large regions of DNA-like fibrils, in the growing filament cells of the green alga Halimeda. × 28 500. (Dr M.A. Borowitzka, University of Sydney.)

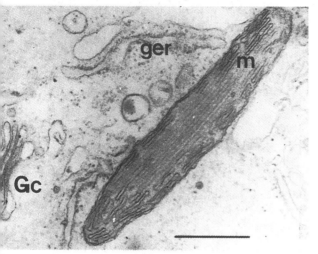

Fig. 57(b) – Growing filament cells of the green alga *Halimeda* with a mitochondrion (m) containing striated bands (DNA) running along its length. × 32 000. (Dr M.A. Borowitzka.)

Right
Fig. 58(a) – An opercular wall cell of the polychaete, *Spirorbis spirorbis*, revealing the spatial arrangement between a Golgi complex (Gc), multivesicular bodies (mvb) and lysosomes (ly). × 20 000. (Dr A. Bubel.)

The shape of mitochondria is variable but in general the organelles are rod- (or sausage-) shaped, spherical and ovoid. Other forms may be seen during certain functional conditions. For example, several rounded mitochondria of the spermatids of polychaete, *Questia* are transformed (possibly by fusion or elimination) into two elongated mitochondrial derivatives, which in transverse section form two crescents enclosing the axoneme (Fig 170a, b). Also, giant mitochondria arise as localized areas of exceedingly long, sinuous and complex-branched mitochondria (Fig 48a–d) in human endometrial glandular cells around the time of ovulation. Frequently, flattened, pancake- or plate-shaped mitochondria are observed in the gills of the blue crab, *Callinectes sapidus*, and in the pulmonary cells of the laboratory-reared newt, *Triturus alpestris*.

The size of mitochondria is also variable in most cells; the width is relatively constant averaging 0.5 μm, whereas the length ranges between 2 and 7 μm. In some cells, giant mitochondria coexist with a population of normal-sized ones. Giant mitochondria are localized swellings (up to 1.5 μm in diameter), that occur in exceedingly long, sinuous, complex-branched mitochondria and are found in human endometrial glandular cells among normal-sized mitochondria (0.3 μm diameter). Two types of mitochondria are found in mouse liver cells; namely, small ones about 0.4 × 1.0 μm and larger ones up to 9 μm in length (Fig 49a, b). The presence of giant mitochondria is reported in addition in a number of other situations, such as in the cells of yeasts, protozoa, insects and cardiac muscle and liver cells of higher animals. In some algae, giant mitochondria are a feature of their development cycle and coincide with a decrease in respiratory rate.

Mitochondria are, in general, uniformly distributed throughout the cytoplasm, but there are many exceptions to this rule. In some cases they accumulate preferentially around the nucleus or in the peripheral cytoplasm. Their distribution within the cytoplasm should be considered in relation to their function as energy suppliers. For example, in certain muscle cells (e.g. diaphragm), mitochondria are grouped like rings or braces around the I-band of the myofibrils. In rod and cone cells of the retina, all mitochondria are located in a portion of the inner segment. In kidney tubule

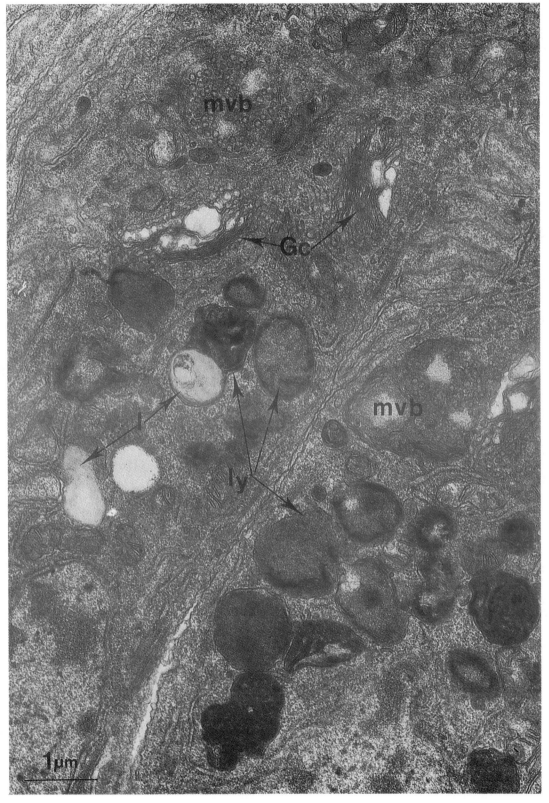

1µm

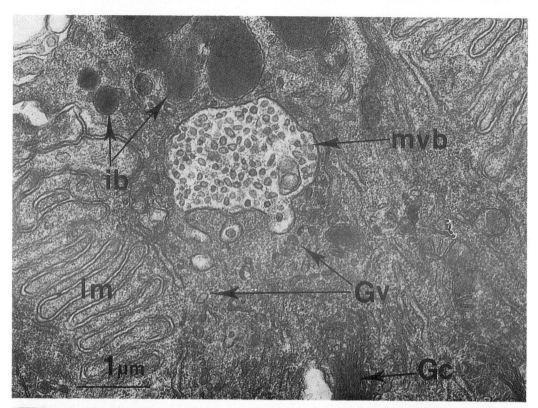

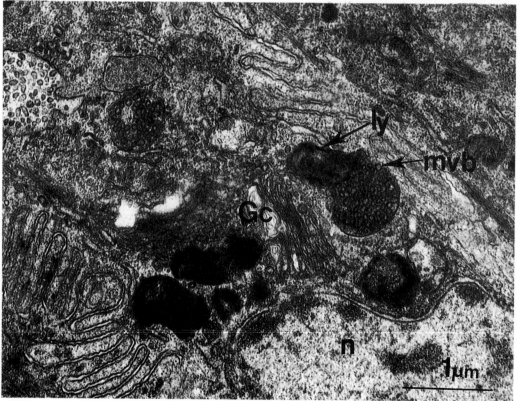

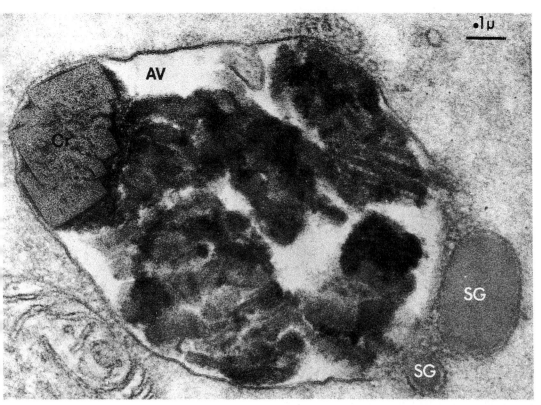

Above left

Fig. 58(b) – A small portion of an opercular wall cell of *S. spirorbis* showing the close association between Golgi vesicles (GV) and multivesicular bodies (mvb). × 19 000. (Dr A. Bubel.) ib, inclusion bodies.

Below left

Fig. 58(c) – The supranuclear region of an opercular wall cell of *S. spirorbis* showing the apparent fusion of a multivesicular body (mvb) and a lysosome (ly). × 25 000. (Dr A. Bubel.)

Above

Fig. 59(a) – An autophagic vacuole (AV) containing a crystalline substance (Cr) as well as amorphous and membranous material in an accessory boring gland cell of the gastropod, *Urosalpinx*. There are also two secretory granules (sg) in close proximity to the autophagic vacuole. × 105 000. (Dr P. Nylen.)

cells of many higher animals and in the gills of many invertebrates, mitochondria are intimately associated with infoldings of the plasma membrane. However, in the cells of the distal segment of the kidney of the lamprey and in the chloride cells of the teleost *Cyprinodon variegatus* gill, mitochondria are intimately related to a three-dimensional branched network of cytoplasmic tubules (Figs 104a, 105a). Similarly, in the primary cells of the Malpighian tubules of the larvae of the saline-water mosquito, long microvilli (3–4 μm in length) contain mitochondria along their length (Fig 21a–d). It is assumed in the above examples that the association of mitochondria with a well-developed membrane system is related to the supply of energy for the active transport of water and solutes. Interestingly, during cell division (mitosis) mitochondria are concentrated near the spindle, and upon division of the cell they are distributed in approximately equal numbers between daughter cells.

The number of mitochondria in a cell is difficult to determine, but, in general, it varies with the cell type and its functional state. It is estimated that a normal liver cell contains between 1000 and 1600 mitochondria. The

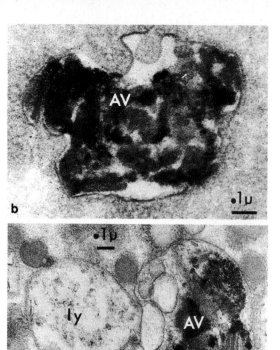

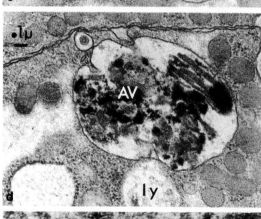

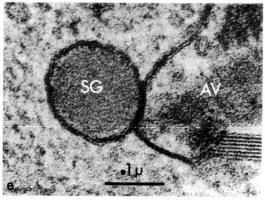

largest number of mitochondria recorded is 300 000 in some oocytes. In cancer cells the number of mitochondria decrease. This decrease is believed to be related to an increase in anaerobic glycolysis in the cells. Generally there are fewer mitochondria in green plant cells than in animal cells since some of their functions are taken over by chloroplasts. There is also a scarcity of mitochondria in undifferentiated cells (Figs 148a, 151a), lymphocytes and cells of the epidermis. Although fibroblasts and secretory cells have an average number of mitochondria, mitochondria are extremely numerous in the liver, parenchymal cells, parietal cells of the gastric gland, proximal kidney tubule cells and cells of the adrenal cortex.

Mitochondria may have a more or less definite orientation. In secretory columnar cells they are generally orientated in an apicobasal direction, parallel to the main axis. In leukocytes, mitochondria are arranged radially with respect to centrioles. It has been suggested that mitochondrial orientation is dependent upon the direction of diffusion currents within cells and is related to the submicroscopic organization of the cytoplasmic matrix and vacuolar system.

Detailed structural variations are also evident in cristae. Cristae may be membranous (flat), tubular or vesicular. In nerve and striated muscle, cristae may be arranged longitudinally or they may be simple or branched, forming complex networks. In protozoa, insect and adrenal cells of the glomerular zone, cristae may be tubular instead of lamellar, and they may be packed in a regular fashion. In the secretory cells of the gastropod accessory boring organ, mitochondria possess straight or wavy cristae that are aligned parallel to the long axes of the organelles (Fig 50). However, similarly orientated cristae in the mitochondria of newt pulmonary cells possess septae (Fig 51a–d). In contrast, the mitochondria in the muscle of the crayfish, *Procambaraus clarkii*, possess triangular or rhomboidal cristae sur-

Fig. 59(b) – Autophagic vacuole (AV) in secretory boring gland cells of *Urosalpinx*, containing membranous arrays. × 63 000. (Dr P. Nylen.)

Fig. 59(c)–(e) – Autophagic vacuole (AV) in secretory boring gland cell of *Urosalpinx* incorporating cytoplasm, lysosomal bodies (ly), vacuoles and secretory granules (sg). 59(c) × 42 000. 59(d). × 32 000. 59(e) × 157 000. (Dr P. Nylen.)

rounded by an array of filaments (Fig 52).

A peculiar feature of the receptor cells (cerebral ocelli) in the serpulid *Spirobranchus giganteus* is the presence of numerous striated fibres, identical in structure and banding periodicity to ciliary rootlets. The rootlets run closely juxtaposed to one or more longitudinally orientated mitochondria. The cristae are massed on the surface adjacent to the rootlets and display a regular spacing closely equivalent to the banding periodicity (Fig 53).

The number of cristae per unit volume of a mitochondria is variable. Mitochondria in liver and germinal cells have few cristae and an abundance of matrix, whereas those in certain muscle cells have numerous cristae and little matrix. The greatest concentration of cristae is found in the flight muscle of insects (Fig 117). In general there seems to be a correlation between the number of cristae and the oxidative activity of the mitochondria. Closely packed membranous cristae characterize mitochondria with a high rate of oxidative metabolism (skeletal muscle, proximal and distal tubules of the kidney) (Fig 104a). Tubular and vesicular cristae are present in cells which synthesize steroid hormones (Fig 54).

Although respiration and oxidative phosphorylation are the most important functions of mitochondria, another related function is the accumulation of cations. In cardiac muscle cells (adult mice/guinea pigs) mitochondria are frequently closely juxtaposed to myocardial gap junctions (Fig 9). Such mitochondria are known to sequester calcium ions and may function to buffer the intracellular calcium ion concentration near the gap junctions. In the matrix of many mitochondria are highly electron-dense intramitochondrial granules, on average 50 nm in diameter, that may represent binding sites for calcium ions. Crystal formation is observed to start in such granules, in some cell types, during pathological calcification. Inclusions such as protein crystals, glycogen particles and myelin (lipid) figures can be present in some mitochondria, also (Fig 151c). Tube-like structures 20–35 nm in diameter arranged in a monolayer are found in the matrix of mitochondria in the tunicate *Pyura vittala,* the cells of the renal tubules of the hibernating snake, *Elphae quadrivirgala,* and in pancreatic cells of the lizard, *Varanus nilotus.* The tubules in the tunicate *P.vittala* are filled with an electron-dense core and possess

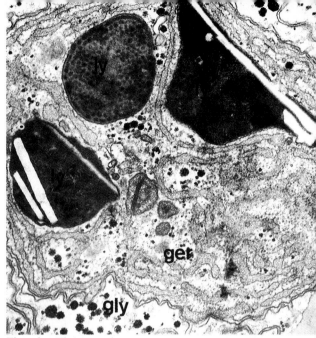

Fig. 60 – The central region of the gonad epithelial cells of the sea anemone, *Actinia fragacea,* containing lysosomes (ly) with their characteristic angular electron-lucent inclusions. × 27 000. (Dr A. Larkman.)

a fairly regular banding pattern of 7 nm width (Fig 55a, b). There are from 14 to 16 such tubules per mitochondria and they are believed to be involved in mitochondrial elongation. In contrast, in the lizard *V.niloticus* the tubules are polygonal and grouped in bundles with 9–10 nm thick filaments. The filaments are arranged in parallel layers, crossing each other at an angle of 57° (Fig 56). In rare instances, microcylinders are also found in the dilated cristae of mitochondria in the pinealocytes of the rat. The tubules possess a dense core and are arranged parallel to one another and to the long axis of the mitochondria.

Mitochondria contain DNA molecules and ribosomes and may synthesize proteins. Furthermore, mitochondria are self replicating. They undergo division and may carry biological information which represents a type of cytoplasmic inheritance. In the matrix of the mitochondria of the alga *Halimeda* DNA appears as delicate strands and crystalline bands about 33 nm wide running the length of the organelle (Fig 57a, b).

Two mechanisms for the biogenesis of mitochondria are proposed, namely: by division from a parent organelle or *de novo* from simpler building blocks.

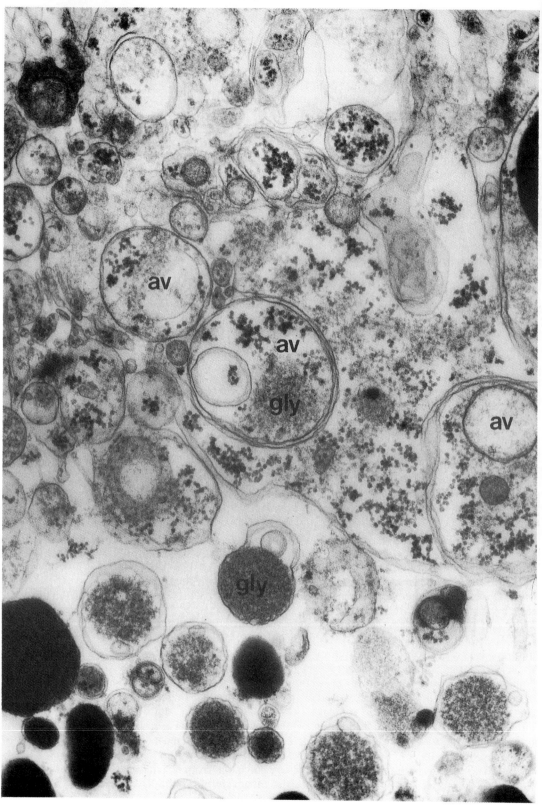

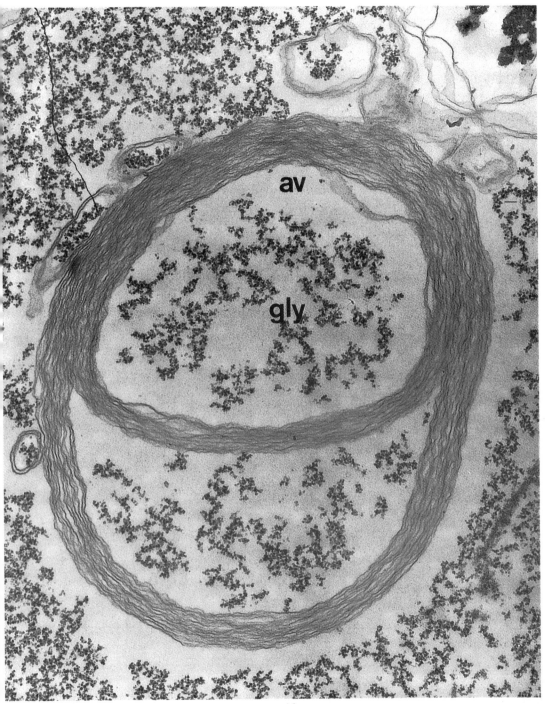

Left
Fig. 61(a) – The sequestration of ß-glycogen (gly) particles into single and double membrane-bound vacuoles (autophagosomes/autophagic vacuoles) (av) in an adipogranular cell from the mantle of the mollusc *Mytilus edulis*. × 47 000. (Dr A. Bubel.)

Above
Fig. 61(b) – A portion of the cytoplasm of a vesicular connective tissue cell from the mantle of *M. edulis* containing large packages of ß-glycogen particles (gly) within autophagosomes/autophagic vacuoles (av) bounded by concentrically arranged membranes. × 53 000. (Dr A. Bubel.)

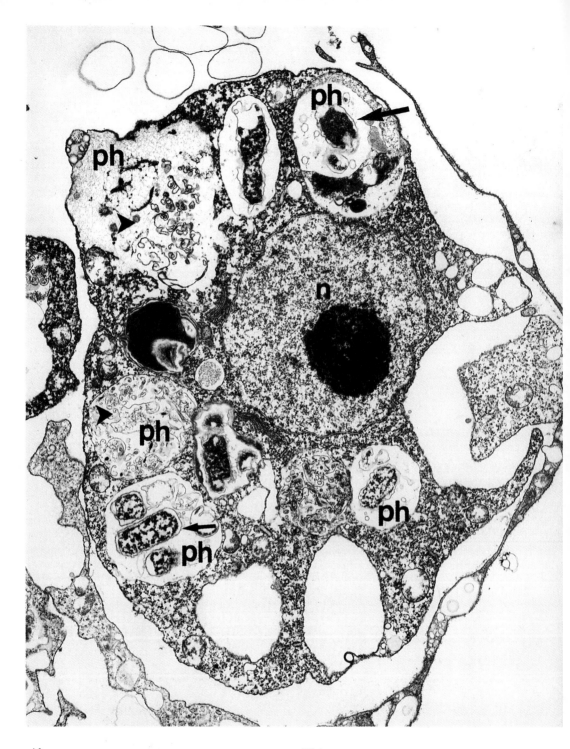

Above
Fig. 62 – A choanocyte of the sponge, *Ephydatia fluviatilis* with phagosomes (ph) containing altered bacteria *E. coli* (arrows) and membranous debris. × 5 000. (Dr Ph. Willenz, Yale University.)

Right
Fig. 63(a) – A diagram of the sequence of events involved in the pinocytosis of material by endothelial cells lining rat liver.

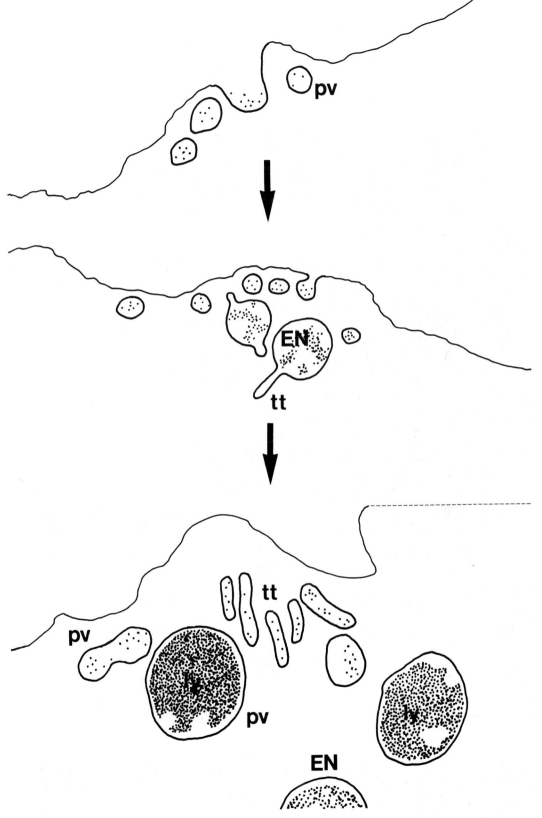

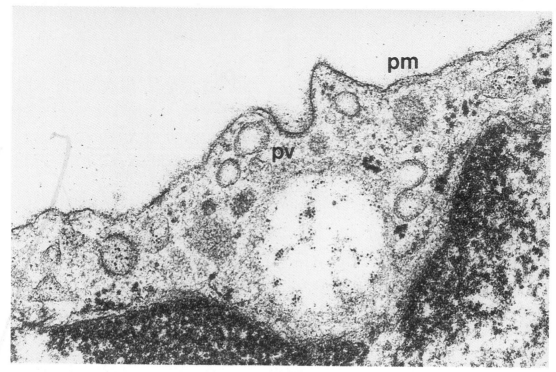

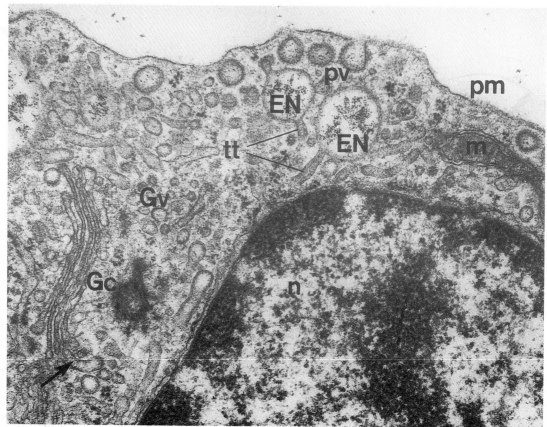

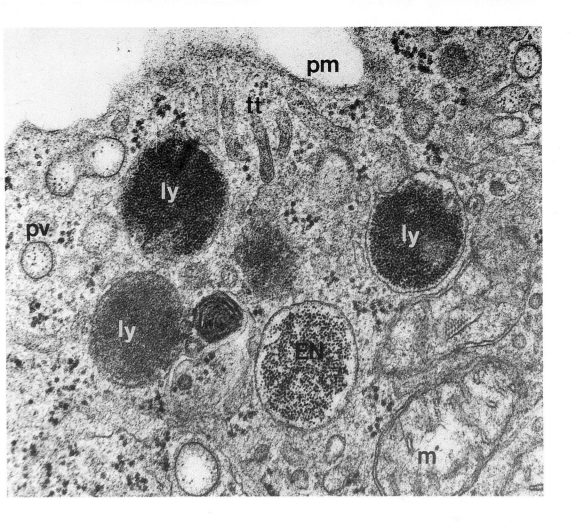

LYSOSOMES

Lysosomes are membrane-bound organelles which contain lytic enzymes for intracellular digestion. They represent an essential part of the cells defence and transport system and abound in cells such as macrophages. Despite the fact that lysosomes are known to contain at least 50 acid hydrolases, including various phosphatases, nucleases, glycosidases, proteases, peptidases, sulphatases and lipases, the most widely used marker enzyme is still acid phosphatase. Lysosomes are spherical or ovoid organelles which range from 25 nm to 0.8 μm in diameter and are present in most cell types. They are bordered by a 6 nm-thick membrane and contain a multitude of varied membraneous and granular structures (Figs 58, 59, 60).

Structurally and functionally, lysosomes are divided into two categories (a) primary and (b)

Left above

Fig. 63(b) – The adluminal surface of rat liver sinus endothelial cells with several bristle-coated pits and bristle-coated vesicles (pv) in the cytoplasm. × 23 000. (Dr P.P.H. de Bruijn, University of Chicago.)

Left below

Fig. 63(c) – Ferritin in bristle-coated pits and bristle-coated vesicles (pv) in rat liver sinus endothelial cells. Two endosomes (EN) containing ferritin are connected with transfer tubules (tt). One cisternae of the Golgi complex continues laterally (arrow) into a wider tubule with the luminal calibre of a transfer tubule. × 50 000. (Dr P.P.H. de Bruijn).

Above

Fig. 63(d) – Endothelial cell cytoplasm, from the liver sinus, containing bristle-coated vesicles (pv) endosomes (EN), transfer tubules (tt) and lysosomes (ly) containing ferritin. × 75 000. (Dr P.P.H. de Bruijn.)

95

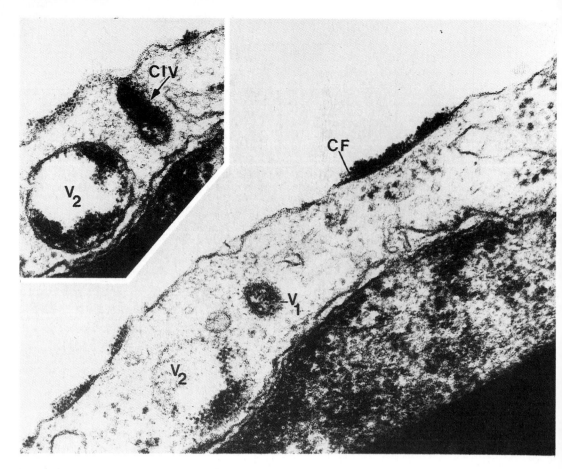

secondary. Primary lysosomes are organelles which have not yet become engaged in enzymatic digestive activities. They are identical to Golgi vesicles and range from 24 to 50 nm in diameter. The enzymes of the lysosomes are believed to be synthesized at the ribosomes of the granular ER, shed into the cisternae, and subsequently transferred to the Golgi apparatus for packaging into small vesicles. Primary lysosomes function during pinocytosis, phagocytosis and autophagy. The fusion between phagosome and primary lysosome results in a merged organelle, referred to as a secondary lysosome. During autophagy (self-phagocytosis), part of the cytoplasm becomes sequestered by a confluence of vesicular or flat paired membranes, derived from either the Golgi apparatus or the agranular ER. Primary lysosomes are added to this structure to form a secondary lysosome (Figs 59a–c, 61a, b). Secondary lysosomes are known by a variety of names: multivesicular bodies, residual bodies, autosomes, haemosiderin and lipofuschin

Above
Fig. 64(a) – A small portion of a rat ovarian granulosa cell incubated for 5 minutes cationized ferritin (CF). A coated membrane invagination (CIV) and cytoplasmic vesicles (V_1 and V_2) contain cationized ferritin. × 120 000. (Dr E. Anderson, Harvard Medical School.)

Right above
Fig. 64(b) – A rat ovarian granulosa cell incubated in a medium containing methylamine and cationized ferritin for 90 minutes. Large cationized ferritin-containing vesicles (V) are surrounded by smaller ones. × 76 000. (Dr E. Anderson.)

Right below
Fig. 64(c) – A rat ovarian granulosa cell incubated for 60 minutes with cationized ferritin, with a multivesicular body (MVB) containing areas of cationized ferritin. × 120 000. (Dr E. Anderson.)

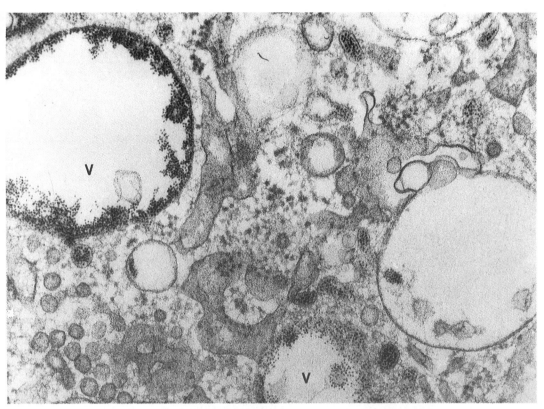

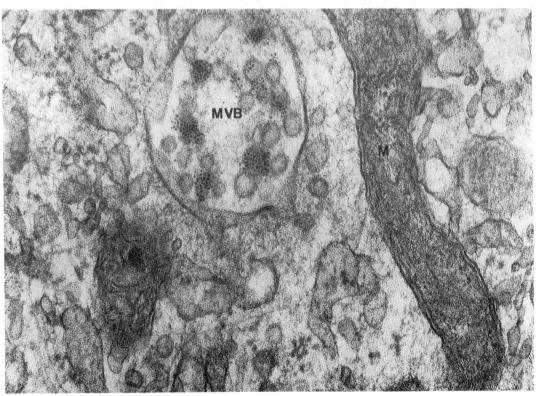

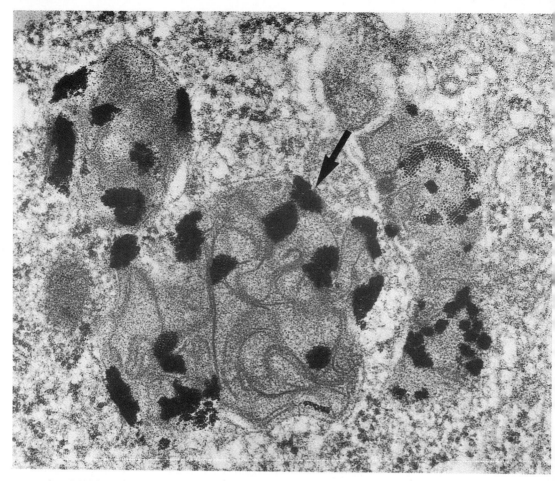

granules. Multivesicular bodies are membrane-bound vesicles which contain small vesicles (Fig 58a–c). However, the role of multi-vesicular bodies as lysosomes remains questionable. Autophagic vacuoles (auto-somes; cytolysomes) are secondary lyso-somes containing, usually, a variety of cell organelles and membraneous bodies. They dispose of and digest damaged, unwanted or aged components of the cytoplasm. In active secretory cells, lysosomes and autophagic vacuoles appear to provide a means for regu-lating oversecretion. Products of hydrolysis taking place in secondary lysosomes diffuse out into the cytoplasm, and there they may be re-utilized. The indigestible residue, after lysomal action, gives rise to residual bodies. Lipofuschin granules represent secondary lyso-somes which accumulate in ageing cells. They contain a variety of lipid droplets, vacuoles and electron-dense granules that give cells a brownish yellow colour.

Above
Fig. 64(d) – Cationized ferritin-containing vesicles in rat ovarian granulosa, possessing membranous material and strong acid phosphatase activity (arrow). × 80 000. (Dr E. Anderson.)

Right
Fig. 65(a) – A haptocilium of the turbellarian *Paratomella rubra*. The shaft is composed of two distinct regions, a long proximal section and a short, relatively thin, distal section which bears the tip. Between the two, where the diameter of the shaft changes, is an abrupt shelf (arrow). At the region of the shelf, four of the peripheral doublet fibrils drop out so that the distal region of a haptocilium has seven fibrils (five doublets plus a pair of singlets). Inside the haptocilium tip is a lamellar, electron-dense core, and outside the ciliary membrane of the tip is an adhesive material. × 37 000. (Dr S. Tyler, University of Maine.)

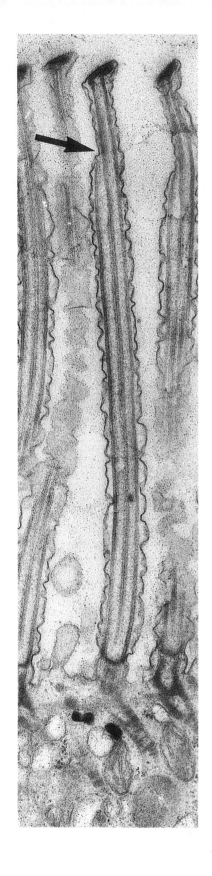

PINOCYTOSIS (ENDOCYTOSIS) AND PHAGOCYTOSIS

The terms pinocytosis and phagocytosis mean, 'cell drinking' and 'cell eating', respectively. The processes participate in the uptake of extracellular materials (endocytosis), such as proteins, metabolic end-products, foreign particles and bacteria (Fig 62). Pinocytotic vesicles result from drop-like invaginations of the plasma membrane. Specific proteins, including clathrin, become associated with the cytoplasmic face of the membrane at these points, and may play a cytoskeletal role in curving the membranes inwards to form invaginations. The invaginations then separate from the plasma membrane to form coated vesicles. The endothelial cells lining the venous tissues of the liver, and the steroid-secreting granulosa cells, pinocytose protein tracers, horseradish peroxidase and ferritin. Pinocytotic vesicles break off from the plasma membrane to form small vesicles between 90 and 120 nm in diameter that contain the tracers (Fig 63a–d). The pinocytotic vesicles may enlarge by fusing with one another to form larger vesicles/vacuoles or endosomes (phagosomes; phagocytotic vacuoles; pinosomes; receptosomes). The vesicles, in some cells, may also fuse with tubules, termed transfer tubules, believed to be derived from a Golgi complex. These subsequently fuse with primary lysosomes, and in some cells with multivesicular bodies, acquire lysosomal hydrolases and become secondary lysosomes (Fig 64a–d). In endothelial cells, hydrolytic enzymes are transferred to endosomes by transfer tubules (Fig 63c). Pinocytotic vesicles also participate in the segregation or discharge of intracellular material by means of exocytosis, or reverse pinocytosis. The origin of vesicles which ultimately end up fusing with the plasma membrane in the process of exocytosis is most often the Golgi zone.

CILIA (FLAGELLA)

Cilia and flagella are cylindrical processes projecting from the surface of epithelial cells. A flagellum is essentially a long cilium. Both are adapted for action in a liquid medium. Although they possess the same basic structure, they differ in their mode of beating and

99

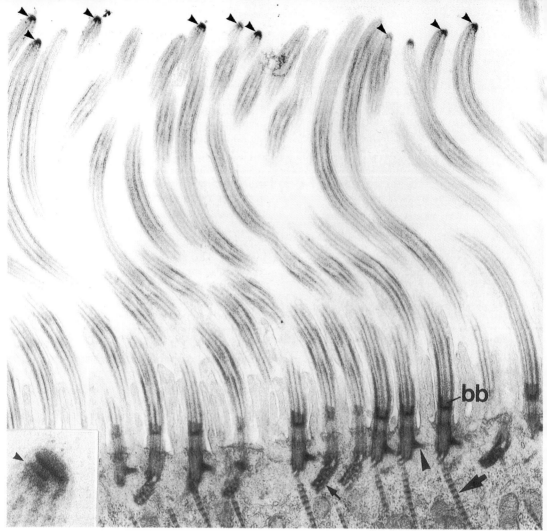

Above

Fig. 65(b) – The cilia of the palate of the frog *Bombina orientalis*. Electron-dense caps (small arrowheads and inset) at the distal tip of the cilia are assymmetrical and all caps are orientated in the same direction across the entire palate. Basal bodies (bb) are anchored to the cell by a basal foot (large arrowhead) and a rootlet complex composed of one long striated (large arrow) and one short rootlet (small arrow). The basal foot points in the direction of the effective ciliary stroke and the smaller cap (arrowhead-inset) points in the opposite direction. × 27 000 (inset × 105 000). (Dr E.L. Lecluyse, University of Kansas.)

Right & centre

Fig. 65(c & d) – Tips of palate cilia of *B. orientalis* showing the attachment of the caps to microtubules as well as to the ciliary membrane and extraciliary crown. The crown (Cr) is composed of short hairs arranged in longitudinal rows across the external face of the ciliary membrane (c, d1) and is attached to the

larger cap subjacent to the membrane. The large upper cap (U) is disc shaped and covers the entire cross-section of the distal tip (d2). Cross-sections (d3–5) reveal that the A microtubules of doublets 4–7 are linked to the large cap (c). The smaller cap (L) lies beneath and to one side of the larger one and is linked to the A microtubules of doublets 1–3, 8 and 9 and the central pair of microtubules (d3).

Similar to mammalian trachael cilia, the B microtubules of each doublet terminates proximal to the A microtubules (d7). Each of the A microtubules has a dense structure within its lumen at the site of attachment to the caps (arrowheads in d3, 4). In longitudinal section (c) these densities appear as plug structures (P) similar to those found in protozoa and tracheal cilia. × 155 000. (Dr E.L. Lecluyse.)

Far right

Fig. 65(e) – Branchial epithelial cell cilium with distended paddle-like (arrow) apex in the polychaete *Spirorbis spirorbis*. × 11 500. (Dr C.A.L. Fitzsimons.)

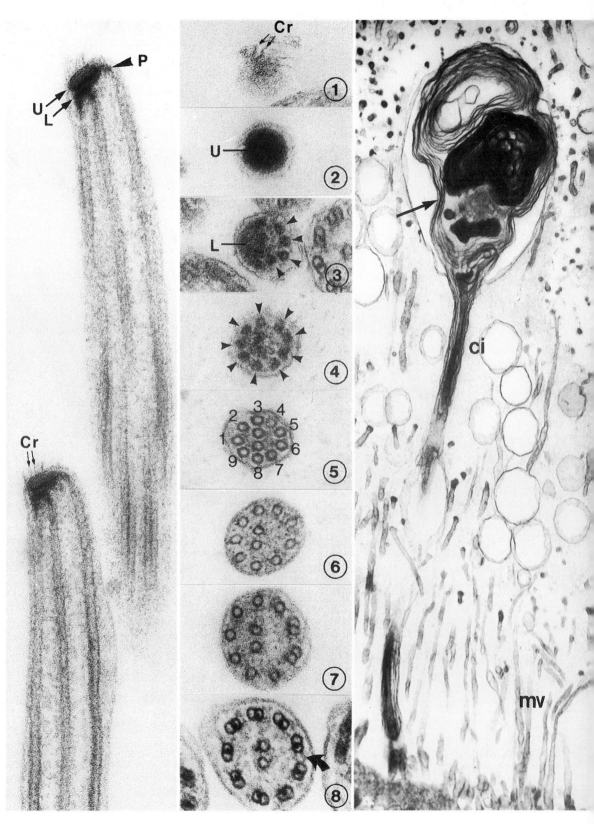

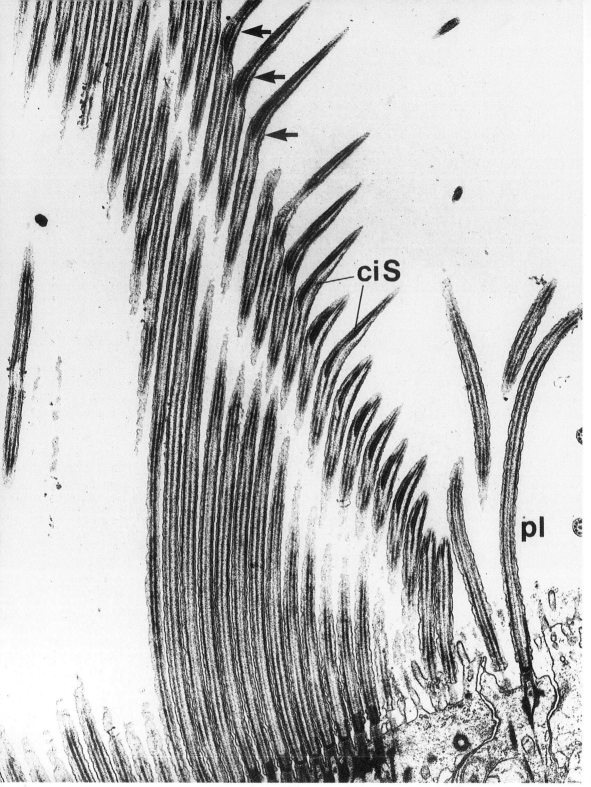

Fig. 66(a) – A longitudinal section through a eulaterofrontal cirrus of the mollusc *Mytilus edulis* showing the marked bend in each cilium (arrows) and the associated stiffening rods (ciS). × 17 000. (Prof. G. Owen, University College of Wales, Aberystwyth.)   pl, prolaterofrontal cilia.

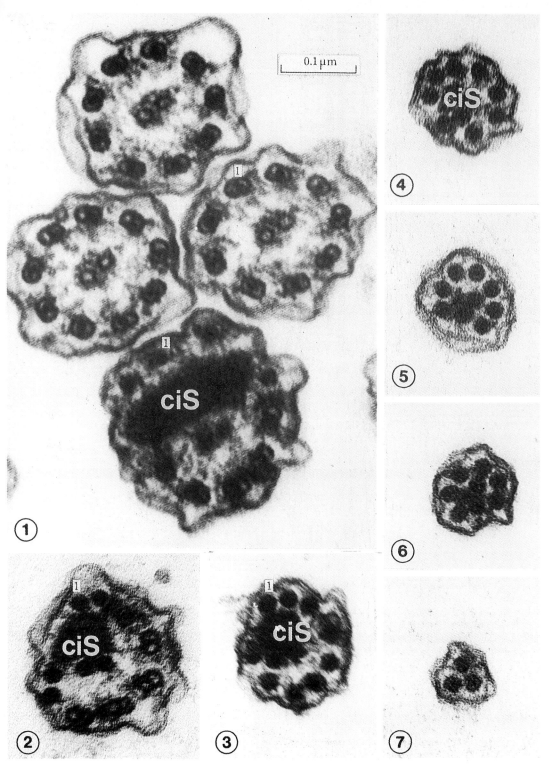

Fig 66(b) – Transverse sections through the cilia of the eulaterofrontal cirrus from the region of the bend to the distal tip (b1–6) and showing the stiffening rods (cis) and changes in the form and number of the ciliary fibrils. × 200 000. (Prof. G. Owen.)

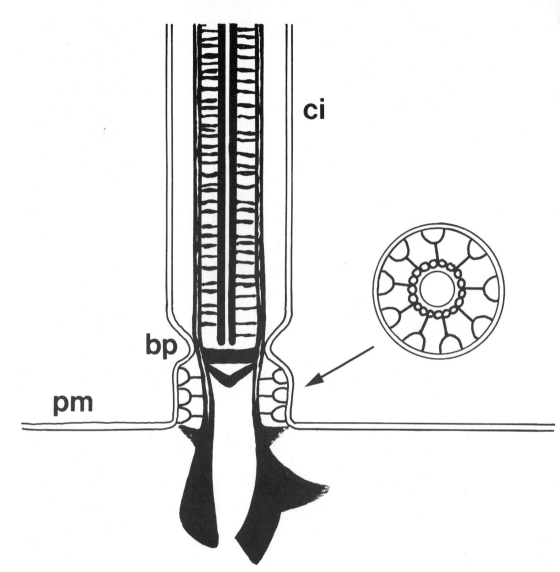

Above

Fig. 67 – A diagrammatic representation of longitudinal and cross sections through the ciliary necklace region of a cilium of the bivalve *Elliptio sp.* The membrane pinches in at the basal plate (bp). Below the basal plate, the necklace region (arrow) contains three bowls or cups (lines which are apposed against the membrane). (After Gilula and Satir 1972.)

Right above

Fig. 68(a) – Longitudinal section through the adoral zone anlage of *E. maggii,* showing short kineties and kinetid development. The kinetids show gradual maturation from the edge to the anlage centre, in the direction of the arrow. Microtubules which are probably basal in origin lie parallel to the kineties in early stages (PM) and become short and oblique postciliary microtubules (PC) while a kinetodesma (KD)

develops in the same direction. Dense material (DM) accumulates at the base of the kinetosomes and centrally in the kineties and at the ends there are short orthogonal prokinetosomes (PK). × 30 000. (Dr D.N. Furness.)

Right below

Fig. 68(b) – Section through an anlage of the entodiniomorphid ciliate, *Endiplodinium maggii* showing kineties of complete kinetids (K_2–K_1). An adjacent kinetosome (K_1) has a different orientation and possesses a probable subcortical kinetid pattern. Basal (B) and cortically directed (C) type microtubules may be present in orientations similar to that of postciliary (PC) and transverse (TM) microtubules, respectively. The kinetosome is orthogonally placed (arrow). × 59 000. (Dr D.N. Furness, University of Keele.)

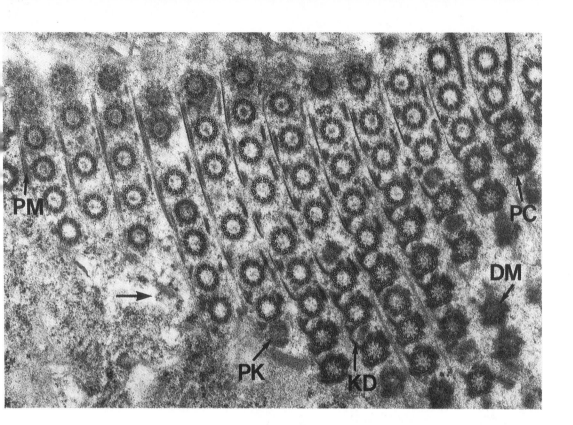

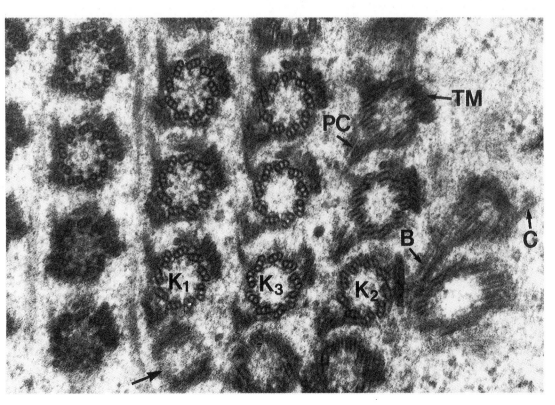

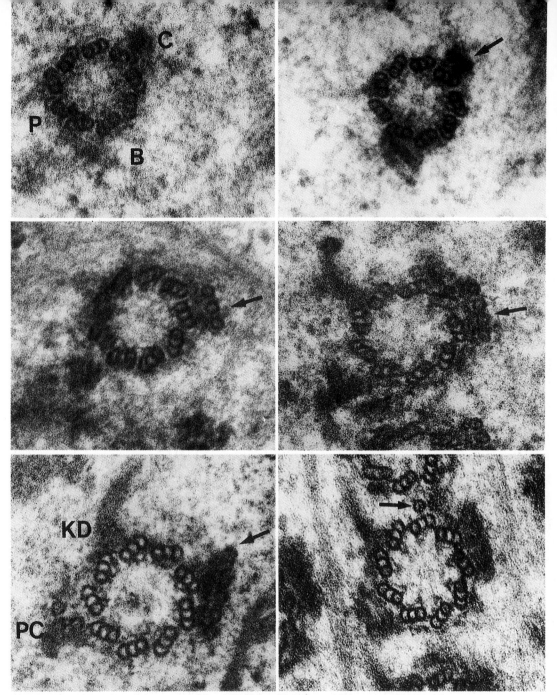

Fig. 69 – Development of the mature kinetids of the ciliary zones from somatic kinetids.

(a) Somatic kinetid with basal (B) and cortically (C) directed microtubule and dense projection (P). × 150 000.

(b) Somatic kinetid with an additional cortically directed microtubule (arrow). × 150 000.

(c) Kinetid with three cortical transverse microtubules (arrow). × 150 000.

(d) Kinetid with four transverse microtubules (arrow) orthogonal to a subcortical kinetosome. × 150 000.

(e) Completed kinetid with five transverse (arrow) and three post-ciliary (PC) microtubules in the same orientation as the cortical and basal microtubules of the somatic kinetid. The origin of the kinetodesma (KD) coincides with the dense projections shown in (a). × 150 000.

(f) A kinetid with a short-lived, single anterior microtubule (arrow). × 150 000. (Dr D.N. Furness).

the type of movement that it produces. The beat of a cilium is asymmetrical; it has an active phase, during which movement is brought about, and a recovery phase, which produces no movement. In contrast, the beat of a flagellum is often symmetrical; it is able to bring about movement continuously throughout its beat. Further, cilia are typically much more densely massed than flagella and their beats are coordinated in the well-known metachronal pattern. The forces set up by cilia and flagella are small and are totally inadequate for the production of movement in large animals. They do, nevertheless, serve some locomotory function in protozoans, ctenophorans, platyhelminthes, turbellarians and molluscs. The coordinated beat of cilia is used to assist in feeding, respiration and transportation by effecting the movement of fluid over surfaces, e.g. for filter feeding in bivalves and ascidians, and on epithelia lining alimentary, respiratory and reproductive tracts in vertebrates. In such epithelia, cilia may have a spatial density of one cilium per μm^2 and show a well-developed metachronism, which is characteristic of well-coordinated cilia. However, where maintenance of such a coordination is not critical, cilia may have a low spatial density of one cilium per 10–56 μm^2. Flagella create feeding currents in sponges and they provide for movement of sperm in most animal groups (and in plants).

Cilia are between 5 and 10 μm long and about 0.5 μm in diameter, although they may

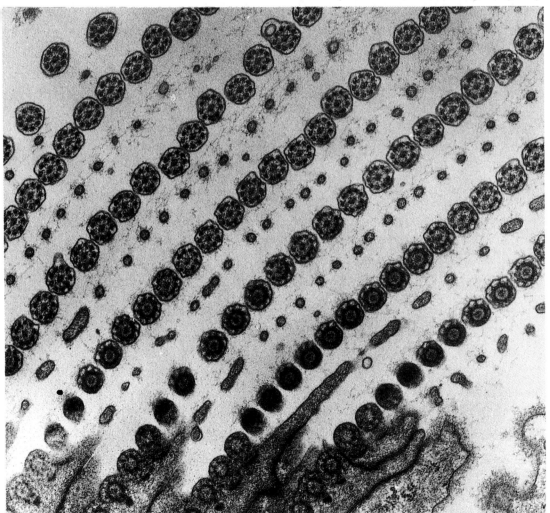

Fig. 70(a) – Oblique section through a branchial stigma of the ascidian, *Botryllus schlosseri*, showing seven rows of cilia and microvilli. × 26.000. (Dr G.B. Martinucci, University of Padua.)

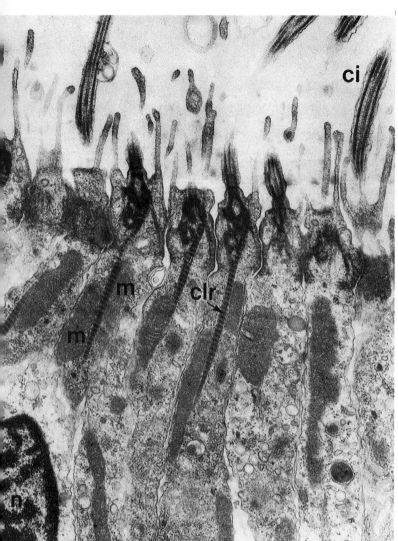

Above

Fig. 70(b) – Stigma cilia (ci) with rootlets (clr) closely associated with mitochondria (m). The mitochondrial cristae appear to be aligned with the rootlet bands. × 24 000. (Dr G.B. Martinucci.)

Right above

Fig. 70(c) – Freeze-fracture replica of a single ciliary row in frontal view. Clusters of IMPs (intramembrane particles) are evident on the P-face of the membrane, distributed in two to three rows and possibly aligned according to the underlying axonemal doublets. The IMP clusters are in the proximal region of the ciliary shaft, but are absent just above the ciliary necklace (arrowheads). × 58 000. (Dr G.B. Martinucci.)

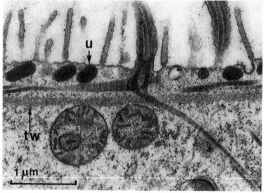

Fig. 70(d) – Cilium of a turbellarian, *Psamnomacrostomum sp.*, rostral rootlet extends towards the lower right, caudal rootlet towards left. × 18 000. (Dr S. Tyler.)

reach between 23 and 25 μm in length in some cells, e.g. myoepithelia of sea anemones. Depending on the epithelium, the number of cilia vary from one to more than 250 per cell, e.g. in respiratory tracts in mammals.

A cilium consists of a ciliary shaft or axoneme composed of a bundle of microtubular fibrils enclosed by a ciliary membrane that is continuous with the cell plasma membrane. Each fibril is formed of a protein called tubulin which resembles actin. The microtubular fibrils are arranged in a 9 + 2 pattern, i.e. nine are doublets, each composed of an A and B subfibril, that form a circle around a central pair of singlets (diameter about 24 nm). Subfibril A in each of the peripheral doublets bears a double row of short arms which all point in the same direction. A plane perpendicular to the line joining the two central singlets divides the axoneme into right and left symmetrical halves. The plane of ciliary beat is perpendicular to this plane of symmetry. The nine pairs of doublets are numbered, starting from the pair lying in the plane of symmetry and continuing in a clockwise direction. Normally, a sheath encloses the central fibrils and delicate strands lie between the central and peripheral fibrils, sometimes forming spokes or radial lamellae.

The central singlets extend to the tip of a cilium whereas subfibrils A and B of the peripheral doublets terminate sequentially below the central singlets. Capping structures, which differ in different organisms, are common at ciliary tips. For example, two caps asymmetrically positioned at ciliary tips are found in acoel turbellarians and amphibian cilia. In the turbellarian cilia, doublets 4–7 end on a shelf-like cap and doublets 1–3, 8 and 9 and the central singlets terminate on a larger cap located distal to the smaller one (Fig 65a). Similarly, in frog palate cilia, doublets 1–3, 8 and 9, and the central singlets terminate in a proximal shelf and doublets 4–7 terminate on a larger distal cap (Fig 65b–d). Such capping structures are thought to be involved in the regulation of microtubular fibril assembly. The caps regulate microtubule length by corking the microtubular fibril ends and so prevent the addition of tubulin. In acoel turbellarians, the tips of cilia are, in addition, specifically modified to serve an adhesive function so as to allow the animals to stick to substrata. The shafts of such cilia or haptocilia are composed of two distinct regions: a long, relatively wide proximal section and a shorter, relatively thin, distal section which bears the adhesive tip. Between the two, where the diameter of the shaft changes, is an abrupt shelf (Fig 65a).

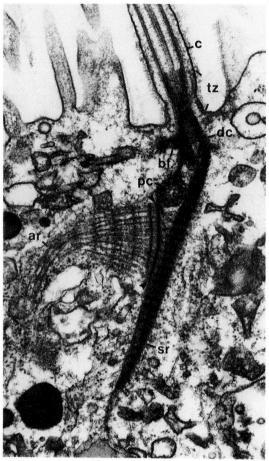

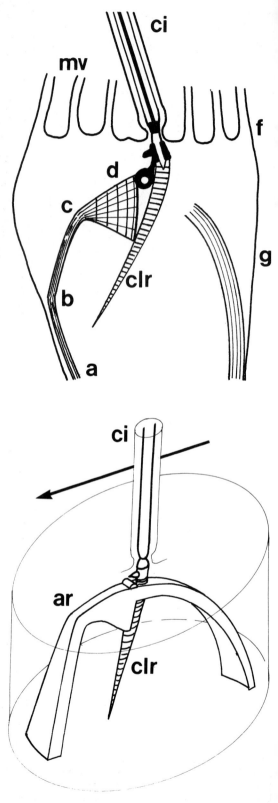

Above left

Fig. 71(a) – Longitudinal section of myoepithelial cell of the sea anemone, *Metridium senile*. × 36 000. (Dr M.C. Holley, University of Bristol.)

ar, anterior part of arched rootlet; bf, basal foot; c, cilium base; dc, distal centriole; pc, proximal centriole; sr, striated rootlet; tz, transition zone.

Above right

Fig. 71(b) – Diagram of myoepithelial cell apex with ciliary basal apparatus sectioned longitudinally in the same plane as the path of the cilium during an effective-stroke. The effective stroke in this diagram is from right to left. Projecting from the cell surface into the coelenteron is a cilium (ci) and numerous microvilli (mv). A septate junction, the upper and lower limits of which are marked 'f' and 'g' respectively, forms a belt around the cell. The striated ciliary rootlet (clr) is flanked by an arched rootlet (a, b, c, d). (Dr M.C. Holley.)

Right

Fig. 71(c) – A three-dimensional diagram of the basal apparatus inside the cell apex (fine line to illustrate the arched rootlet (ar). The arrow indicates the direction of the ciliary effective stroke. (Dr M.C. Holley.)

110

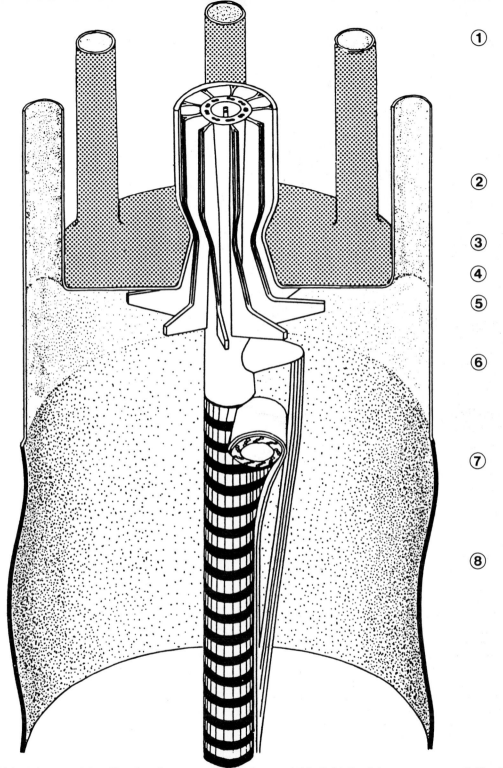

Fig. 72(a) – Diagram of the ciliary basal apparatus in the pharnyx epithelial cells of the sea anemone, *Calliactis parasitica*. The cell cap, bearing microvilli, sits above the belt desmosome, which is indicated by the shaded part of the membrane. The power-stroke of the cilia is left to right across the page in the same plane as the basal foot. The numbers refer to the levels of the transverse sections of the basal apparatus in Fig. 72(b). (Dr M.C. Holley.)

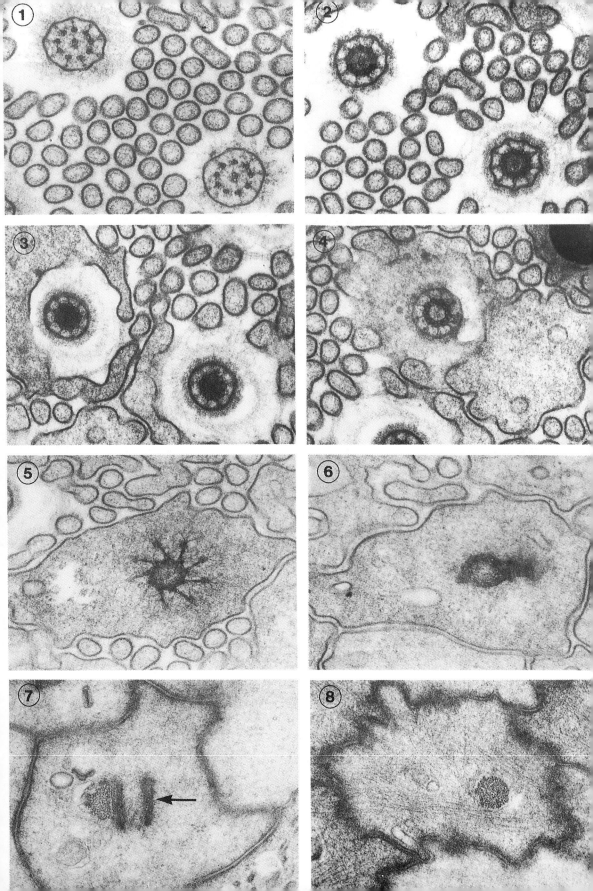

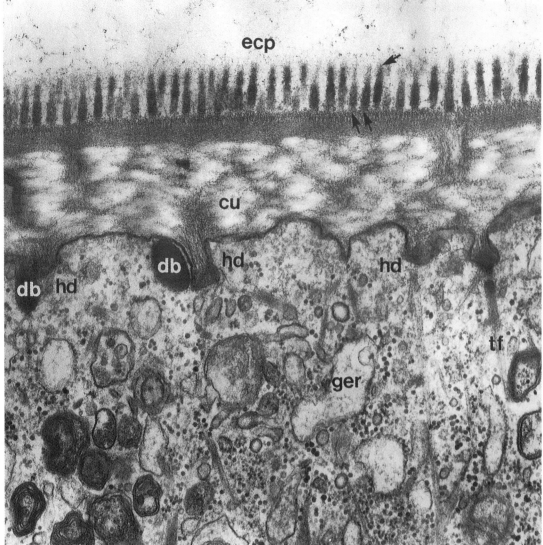

Left

Fig. 72(b) – A sequence of transverse sections through the basal apparatus of pharynx epithelial cilia of *C. parasitica,* as viewed from tip to base. (1) The ciliary axoneme surrounded by microvilli, immediately above the transition zone. (2) The transition zone above the transverse plate. The central pair of microtubules of the axoneme and the radial processes to the ciliary membrane are clearly seen. (3) The transition zone through the transverse plate. The central pair of axonemal microtubules originate from here. (4) The transition zone below the transverse plate. (5) The distal centriole immediately below the cell surface. The processes radiating from the centriole are attached to the cell surface. (6) The distal centriole and the striated basal foot. At this level, the cell membranes are loosely associated with those of the adjacent cells. (7) The proximal centriole (arrow) embedded in the rootlet immediately below the distal centriole. (8) The rootlet immediately below the proximal centriole in the belt desmosome region. In this section, taken from a contracted epithelial cell, filament bundles and cell-membrane buckling are seen. × 46 000. (Dr M.C. Holley.)

Above

Fig. 73(a) – The epidermal cell apex of the annelid *Enchytraeus sp.,* showing hemidesmosomal-like plaques (hd) adjacent to the cuticle (cu). Filamentous and granular material (arrow) is associated with bottle-shaped ellipsoids (ecp) of the cuticle. Just below the ellipsoids the cuticle is faintly striated (double arrows). × 25 000. (Dr J.M. Burke, University of Washington.)

113

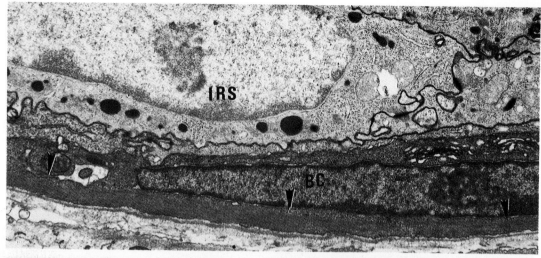

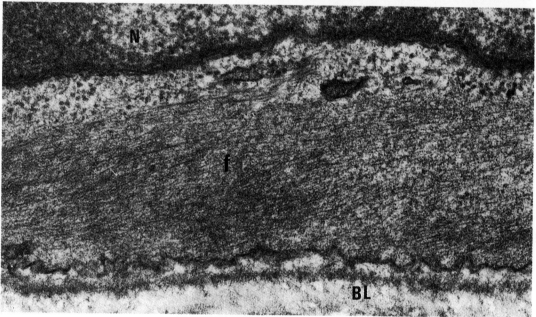

Above top

Fig. 73(b) – A portion of a cross-section through a hair follicle of a mouse. Microfilament bundles (arrowheads) are located in the basal cytoplasm of a flattened basal cell (BC). × 8 000. (Dr S. Sugiyama, Sapporo Medical College.)
IRS, inner root sheath cell.

Above

Fig. 73(c) – Microfilament bundles (f) of the basal cell at the hair bulb. The microfilament bundles are nearly aligned parallel to the basal plasma membrane. × 62 000. (Dr S. Sugiyama.)

Right

Fig. 73(d) – Diagram of Fig. 73(c).

The axonemes in some cilia show further modifications. The eufrontal cirri, which are responsible for the filtering activities of the gills in the bivalve *Mytilus*, are each composed of 22–26 pairs of cilia arranged in two parallel but alternating rows. The distal region of each axoneme is bent to one side or other of the main axis of the cirrus (Fig 66a). The overall effect is to produce a meshwork between the cirri and the gill filament. Basal to the bend, the main body of the cilium shows the typical 9 + 2 pattern (Fig 66b). The bend in each cilium is maintained by a tapering hemispherical structure lying between the peripheral doublet 1 and the central pair of singlets (Fig 66b).

Distal to the bend, all peripheral doublets are modified structurally to singlets and the cilia gradually taper to the tip (Fig 66b). In a number of organisms, namely, annelids, molluscs and protochordates, modifications at the distal ends of cilia appear as biconcave flattened discs or 'paddles'. The axoneme fibril bundle within such paddle-shaped cilia remains intact, except in the distal swelling (Fig 65e). In the protochordate *Rhadopleura* and the bivalve *Mytilus,* such cilia appear to act as minute spatulas for the application of tube and adhesive-disc material, respectively. There is, however, the belief that such paddle-shaped cilia are artifacts. Highly modified, thick, curved, finger-shaped ciliary structures interpreted as 'macrocilia' are found in the epithelia of ctenophores. Each cell bears up to 2000–3000 ciliary axonemes of the 9 + 2 pattern, which are enclosed by a single outer membrane. The axonemes within a macrocilium are packed in a hexagonal array, and neighbouring axonemes are linked by cross-bridges. Mechanical coordination of ciliary activity between successive macrocilia helps in the trapping of prey.

Proximally, just above the junction of the ciliary membrane with the cell plasma membrane, the ciliary membrane constricts to form a tight collar, very closely applied to the peripheral doublets. Near the lower end of the collar, a single or pair of basal plates are found. The central pair of singlets terminate either a short distance from or on the basal plate. In the collar region, between the basal plate and the juncture of ciliary and cell membranes, is the 'ciliary necklace' region. It is characterized by the close packing of peripheral doublets, the absence of central singlets, and by a series of specialized connections that extend from the mid-wall of each doublet to the ciliary membrane. This region is believed to be involved in the control of localized membrane permeability (Fig 67).

Below the collar, the ciliary membrane expands and is continuous with the cell plasma membrane. At this point the axoneme enters the cell as the basal body: an intracellular organelle similar to a centriole. Cilia take their origin in basal bodies. Basal bodies are variously known as basal granules, blepharoplasts and also kinetosomes. The formation of the complex network of cytoskeletal components of the infraciliature in protozoan ciliates is based upon a pre-existing pattern generator in subcortical kinetosomes. Each component that is added during development has its position and orientation predefined (Fig 68a, b, 69a–f). The kinetid (row) pattern of kinetosomes changes sequentially to that of ciliary zones. The basal body is cylindrical and is composed of nine peripheral doublets, to each of which a third microtubular fibril is added. It is, in effect, composed of a 9 + 0 pattern of triplet microtubular fibrils. Proximally, the basal body gives rise to one or more striated

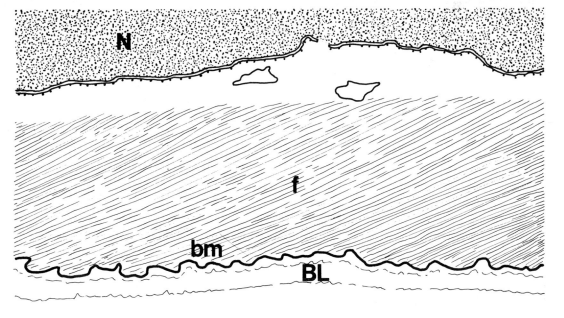

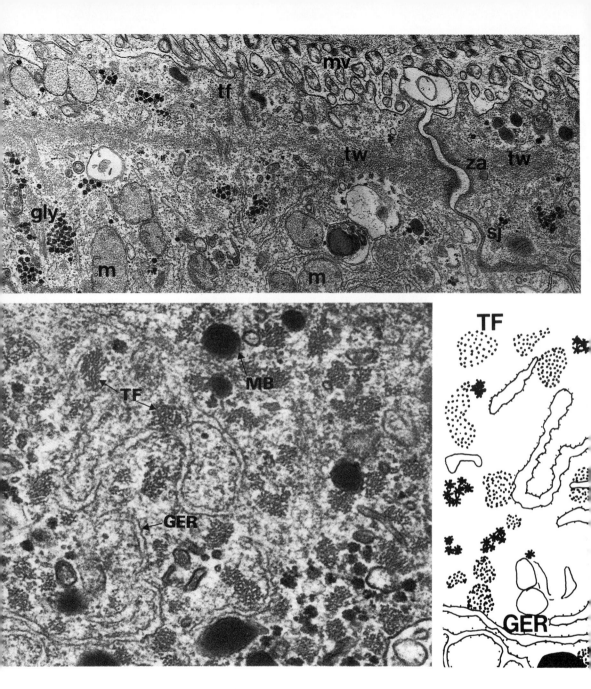

Above left
Fig. 74(a) – The cell apex of a mantle epithelial cell of the bivalve *Mytilus edulis*, showing a well-developed terminal web (tw). × 36 000. (Dr A. Bubel.)

Above right
Fig. 74(b) – Diagram of Fig. 74(a).

Below left
Fig. 74(c) – The apical cytoplasm of an opercular filament epidermal cell of the polychaete *Pomato-ceros lamarckii*, showing the numerous small bundles of filaments. (TF) permeating the cytoplasm. × 62 000. (Dr A. Bubel.)

Below right
Fig. 74(d) – Diagram of 74(c).

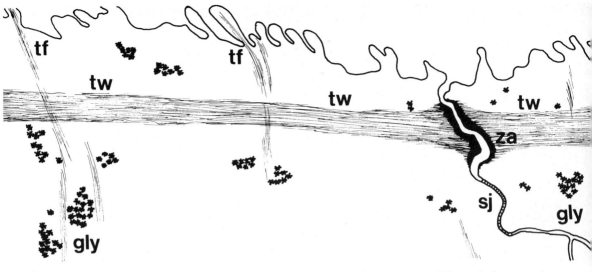

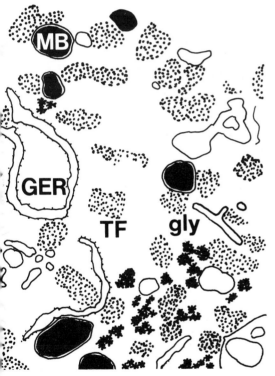

rootlets. The rootlets, according to the location of the cilia, differ in length and in the width and the periodicity of the striations. In addition to the rootlets, structures known as basal feet project from basal bodies.

There is considerable diversity in the structural organization of the basal apparatus in different organisms. The diversity is related to cilia on functionally different epithelia, and implies functions beyond those of axoneme

formation and anchorage. There is increased structural complexity in the basal region of the ciliary shaft, above the ciliary necklace, in cilia lining the branchial sac of ascidians. These cilia are arranged in orderly rows and act in a coordinated manner in a series of metachronal waves (Fig 70a, b). Specializations in these cilia appear as clusters of intramembraneous parti- cles (IMPs) associated with dense material connecting axonemal doublets 5–6 and 9–12 and 2 (Fig 70c). The array of IMPs is located on the opposite faces of each cilium and resemble a kind of ciliary granule plaque. They are thought to be sites for ion chanelling. In epithelia that possess cilia that act in a coordin- ated manner, the cilia have well-developed rootlets and basal feet on their basal bodies. The basal foot always points in the direction of the effective (power) stroke of a cilium. Basal feet on adjacent cilia in the epidermis of the gastropod foot appear to make contact to form a network of interconnected basal bodies. This is thought to play a role in the coordination of their beat. Similarly, locomotory cilia in turbel- larians have a structure that is noteworthy for its pattern of interconnecting ciliary rootlets, with one caudal and one rostral rootlet per cilium (Fig 70d). In ciliate protozoans, patterns of fibres, called kinetodesmata, are closely associated with the basal bodies of cilia. The cilia in these organisms are arranged in rows, each row, or kinety, consisting of a complex which comprises the cilia themselves, their basal bodies and the kinetodesmata which lie associated with each row (Figs 68, 69). A single kinetodesma may be composed of a number

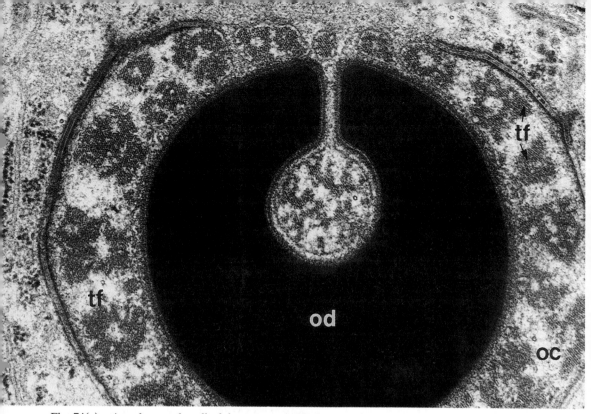

Fig. 74(e) – An odontostyle cell of the nematode *Xiphinem americanum*, with microfilament bundles (tf) in a cytoplasmic sleeve (oc) surrounding an odontostyle (od). × 61 000. (Dr R.F. Carter, Ontario Veterinary College.)

of overlapping fibrils, each of which arises from a basal body. The fibrils are often striated with a periodicity of 40 nm. It has been proposed that kinetodesmata and associated fibrils may be responsible for the coordination that provides for metachronal rhythms in these organisms. In the frog-palate epithelium cilia the rootlet complex is composed of one 3 µm-long and one 0.5 µm-long rootlet. The rootlets are composed of bundles of 16–20 filaments, each 5 nm thick, with cross-striations 30 nm wide at intervals of 70 nm. These rootlets taper until the bundle consists of 4–7 filaments (Fig 65b). Ciliated epithelial cells in coelenterates (sea anemones) differ from most metazoan cilated cells in that their basal apparatus contains a second (proximal) centriole and striated rootlets. The cilium base in monocili-ated pharynx cells of *Calliactis* and myoepithelial cells of *Metridium* is continuous with a basal body (distal centriole) via a transition zone (Fig 71a, b). A single basal foot extends laterally from the distal centriole (Fig 71a). The tip of the foot acts as a focus for many microtubules that radiate out in all directions; some lead to cell surface and lateral membranes in myoepithelial cells. However, in pharynx cells, the distal centriole and microtu-bules are anchored to the cell surface and ciliary membranes by a complex of nine sheet-like processes (Fig 72a–b). Also, a bundle of microtubules, which is attached to the tip of the basal foot, passes to meet a rootlet. In addition, there is a strap-like structure, 2 µm long, which runs parallel to the rootlet on the same side as the basal foot and winds around the proximal centriole. The proximal centriole is situated below the distal centriole, perpendi-cular to it and displaced towards the same side as that from which the basal foot projects. A single rootlet, up to 20 µm long, composed of bundles of 4 nm-thick filaments with striations at a periodicity of 63 nm, is found in the pharynx cells. In contrast, in myoepthelial cells a single striated rootlet up to 2.5 µm long and a structurally different arched rootlet is found (Fig 71a–c). The arched rootlet is composed of 8–11 nm-thick filaments loosely bundled together, and the anterior half, close to the proximal centriole, is striated with a banding period of 140–150 nm (Fig 71a). The anterior and posterior halves of the rootlet are continuous, forming an arch with the vertical axis of the basal apparatus at the top. The rootlet is divided into two at the apex of the arch so that it passes on either side of the short straight central rootlet.

Direct evidence for the function of rootlets is not clearly demonstrated, although they may provide mechanical support for cilia during beating. A structure similar to rootlets, the rhizoplast, in the algal flagellate *Platymonas subcordiformis* is contractile. Interestingly, the presence of contractile proteins and ATPase activity is found in a variety of rootlet structures, e.g. in protozoans, ciliated tracheal cells, human retinal rod cells, and in the flagellar apparatus. However, in cells such as the myoepithelial cells of *Metridium* the arched rootlet is thought to compensate for changes in cell diameter. The two ends of the arched rootlet may in fact function like leaf springs, coming to rest against, and moving away from, the cells' lateral membrane as the cell diameter decreases and increases, respectively (Fig 71c).

CELL WEB (TERMINAL WEB)

An intracellular fibre system that serves in a cytoskeletal capacity is present in a variety of non-muscle cells. The filaments which constitute this system are approximately 6 nm in diameter and may be arranged in a variety of ways. They are most abundant in the stratified squamous epithelial cells of vertebrates and in the so-called 'tendon cells' of invertebrates. Muscles in invertebrates are connected to exoskeletal structures by way of tendon cells. In such cell types, the filaments or tonofilaments occur as large bundles that run parallel to the cells' long axis. The filaments are usually attached to apical and basal hemidesmosomes (Figs 73a–d, 74a–e). The cell web provides such cells with a degree of flexibility as well as rigidity for maintaining cell shape, tensile strength and resilience. Tonofilaments also participate in forming the keratin filaments of vertebrate keratinized cells and in the formation of intracellular hard structures e.g. hooks, in invertebrate cells. Other vertebrate and invertebrate epithelial cells, such as fibroblasts, glial cells and endothelial cells, possess a loose network of filaments. The filaments, in

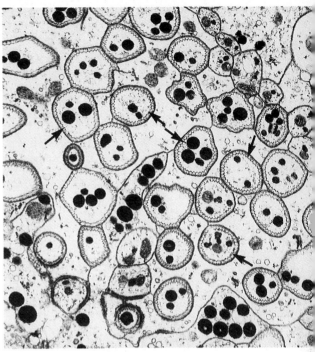

Fig. 75(a) – Transverse section through the mucous cells of the papillae of the tentacles of the polychaete *Magelona sp.* Two circles of microtubules (arrows) alternate around the cells apices. × 15 000. (Prof. Dr V. Storch, University of Kiel.)

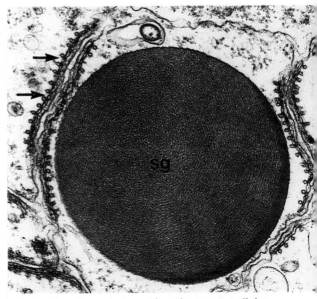

Fig. 75(b) – Transverse section through a cellular process of a collagen gland cell of the bivalve *Mytilus galloprovincialis*. Arrows indicate microtubules associated with subsurface cisternae at the cell apex. × 34 000. (Dr L. Vitellaro-Zuccarello, University of Milan.)

119

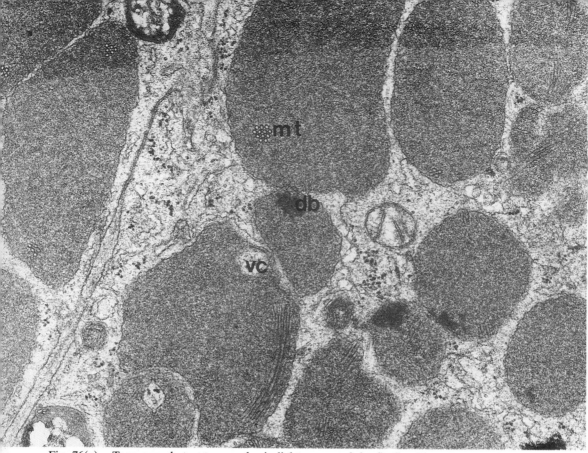

Fig. 76(a) – Type one photocyte granules in light organs of the firefly, *Photuris versicolor,* containing an amorphous matrix, a peripheral dense area (db), flask-shaped vacuole (vc) and bundles of microtubules (mt). × 37 000. (Dr N. Neuwrith, Atlantic Research Laboratory, Nova Scotia.)

addition to providing cytoskeletal support in such cells, play important roles in cell mobility.

At the free surface of epithelial cells with microvilli, immediately below and orientated parallel to the apical plasma membrane, filaments of the cell web are organized into a sheet. The sheet of filaments is called a terminal web (Fig 74a, b). The edges of a terminal web anchor at a zonula adhaerens encircling the apical end of a cell. From the terminal web, filaments may extend into, and form a core of filaments in, microvilli (Figs 19, 20).

Usually, apart from being orientated parallel to the apical plasma membrane, filaments comprising a terminal web are irregularly arranged. Terminal webs may be well developed, either in the form of dense thin sheets, 50 to a few hundred nanometres thick, or as thicker, multilayered sheets from 0.5 to 3 μm thick. Although a terminal web may be absent in some cells, it is most usually seen as a thin band in contact with the apical plasma membrane, making it appear dense. In some invertebrate cells there appears to be an apparent relationship between the development of intercellular junctions and the presence of a terminal web. Cells with a prominent terminal web possess well-developed zonulae adhaerentes and may possess either extensive or relatively inconspicuous septate junctions. In contrast, cells without a terminal web have more extensive septate junctions and zonulae adhaerentes, which may be so inconspicuous so as to be unnoticed.

The main chemical component of filaments of the cell web appears to be actin. It is also evident that at least some of the filaments of a terminal web contain myosin. In some cells, e.g. endodermal epithelial cells, actin filaments are abundant within microvilli cores (Fig 19). The filaments may maintain the shape and provide the motility of such mirovilli.

120

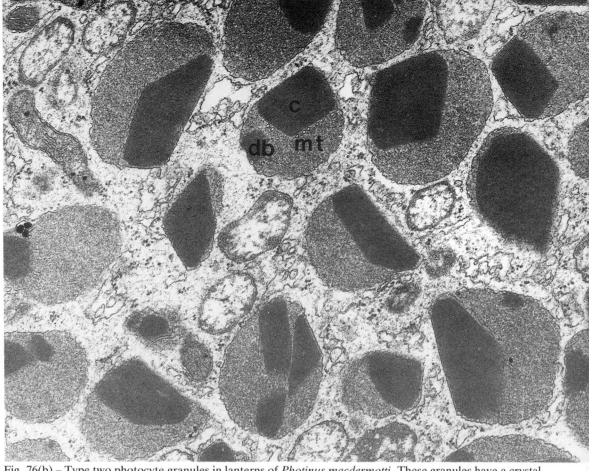

Fig. 76(b) – Type two photocyte granules in lanterns of *Photinus macdermotti*. These granules have a crystal (c) embedded in an amorphous matrix as well as a dense area (db) and microtubules (mt). × 35 000. (Dr N. Neuwrith.)

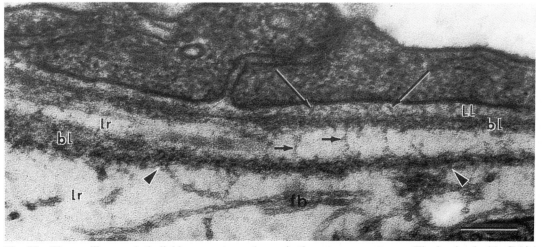

Fig. 77 – The capillary endothelial basement membrane in the rat area postrema. The lumina lucida (LL) is not electron translucent, but is filled with traversing filamentous materials (long arrows). Cross-linker filaments (short arrows) span the two basal laminae (bl). In the outermost lamina reticularis (lr), helical microfibrils (fb) anchor to the outer surface of the basal lamina (arrowheads). × 148 000. (Dr T. Ichimura, Osaka University Medical School.)

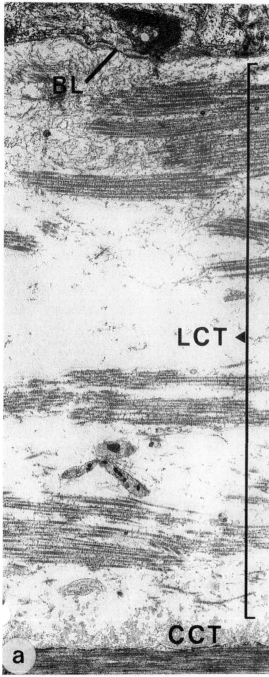

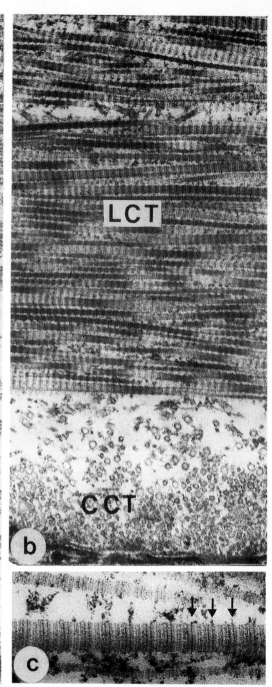

Fig. 78(a) – Longitudinal section through the connective tissue sheath of the tube foot of the sea urchin *Strongylocentrotus franciscanus*. Bundles of collagen fibrils are distributed throughout the sheath. × 108 000. (Prof. Dr E. Florey, University of Konstanz.)

CCT, circular connective tissue layer; LCT, longitudinal orientated layer of connective tissue.

Fig. 78(b) – Part of the connective tissue sheath of sea urchin *Echinus esculentus*, showing densely packed bundles of collagen fibrils. × 29 000. (Prof. Dr E. Florey.)

Fig. 78(c) – A portion of a collagen fibril from the tube foot of *S. franciscanus*. The distance between the major bands (arrows) is 64 nm. × 94 000. (Prof. Dr E. Florey.)

MICROTUBULES

Microtubules are ubiquitous organelles found in the cytoplasm of almost every type of cell. In both animal and plant cells they appear to be similar in nature and dimension. They are on average 25 nm in diameter, bounded by a wall, about 15 nm thick, which encloses an electron-lucent substance, or sometimes a delicate filament. Depending on the cell type, micro-tubules may occur singly, scattered throughout the cytoplasm, in groups arranged in parallel, or fused to form doublets or triplets. These organelles are made up of a protein, the amino acid composition of which closely resembles that of the muscle protein actin.

Microtubules are generally associated with contraction, movement and support in cells. They are believed to serve as a cytoskeleton and also as a means of intracellular transport of materials in the erythrocytes of the frog, lizard, pigeon, goldfish, trout, in mammalian plate-lets, in arthropod haemocytes and in the thin long necks of many gland cells (Fig 75a, b). In the various blood-cell types they occur as marginal bands of between 30 and 40 micro-tubules, whereas in gland cells they are arranged in a circle parallel to the cell's long axis. The arrangement of microtubules in the gland cell necks suggest that they are involved in a secretory granule extrusion mechanism. Unlike other blood-cell types, mammalian erythrocytes do not contain microtubules. Groups of microtubules in nerve-cell axons and dendrites, in the podocytes of renal glomeruli, in crustacean gills and in arthropod Malpighian tubule cells are thought to play a role in intra-cellular transport by serving as diffusion channels for water and other metabolites. However, in fish they facilitate the rapid movement of pigments in melanophores. Structures such as cilia, flagella, sperm tails and the mitotic spindle also contain micro-

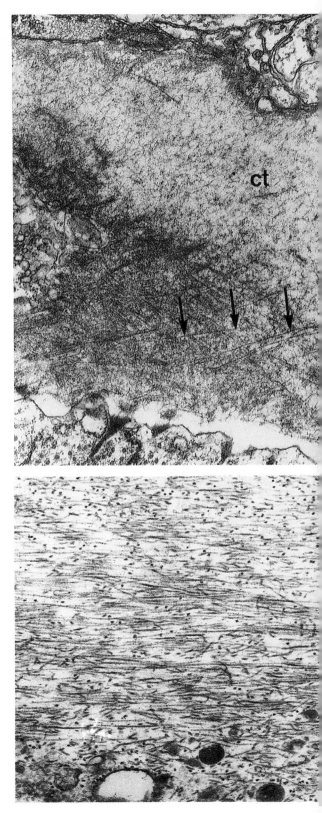

Above right
Fig. 78(d) – The basement lamella which separates the epidermis of the annelid, *Enchytraeus sp.* is composed of fibrils (ct) (arrows) some of which are faintly cross-banded. × 34 000. (Dr J.M. Burke.)

Below right
Fig. 78(e) – Connective tissue fibrils underlying the epidermis of the polychaete, *Pomatoceros lamarkii* peduncle, showing an orthogonal arrangement. × 54 000. (Dr A. Bubel.)

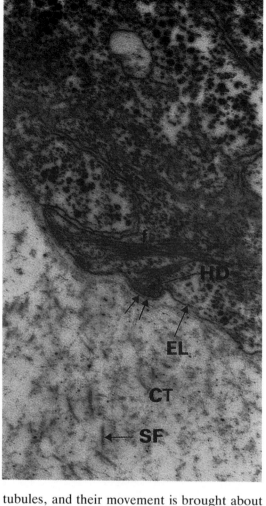

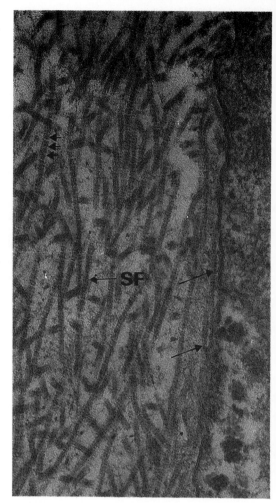

tubules, and their movement is brought about through a sliding mechanism between adjacent microtubules. In addition to playing a role in spindle formation and chromosome movement, bundles of microtubules in some cells are associated with elongated ER cisternae (Figs 28a–c, 40b). They form a continuous system of channels which assures the uninhibited, rapid transport of regulating substances from one part of the spindle to the other and from the poles to the chromosomes or vice versa. Although microtubules are implicated in intracellular movements, such as the change in cell shape and the change in the position of cell organelles during cell division and maturation, they are frequently found associated with other morphogenetic events. For example, microtubules are found parallel to haemocyanin crystals forming in the cyanoblasts of the horseshoe crab, *Limulus*. They are also found

in the photocyte granules of the light organ of fireflies, *Photuris* and *Photinus* (Fig 76a, b). In developing and adult mammalian skeletal muscle, microtubules are predominantly orientated along the long axis of the muscle cells, but in myocardial and rat soleus muscle cells they appear to be helically arranged around myofibrils (Fig 123a–c). Experimental evidence in muscle cells provides support for cytoskeletal and organizational roles for microtubules. When muscle microtubules are disrupted or broken down, skeletal muscle loses shape and the parallel arrangement of the myofibrils is disturbed.

In regions of multicellular organisms where tension is likely to occur, microtubules provide cells with great tensile strength. In invertebrates, for example arthropods, 'tendon cells' are located between muscles and the exoskeleton. The greater part of the cytoplasm of

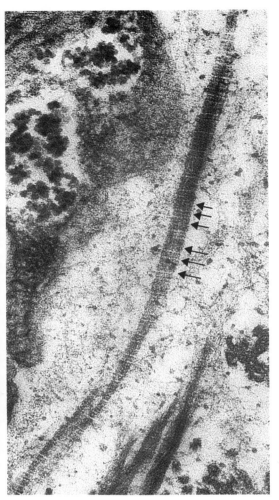

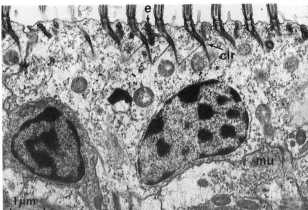

Fig. 79(a) – The simple cuboidal ciliated epithelium of the acoel *Convolulata* sp., which lacks a basement membrane and has an irregular basal surface. Their ciliary rootlets (clr) are interconnected. × 9 500. (Dr S. Tyler.)   e, epitheliosome.

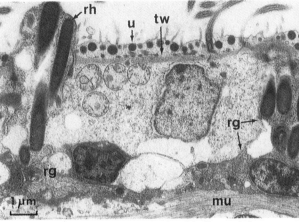

Fig. 79(b) – The simple cuboidal epithelium of the acoel *Psammomacrostomum* sp. which lacks a basal membrane, but has a well-developed terminal web (tw). × 7 000. (Dr S. Tyler.)
rg, rhabdite gland; rh, rhabdite; u, ultrarhabdite.

Above far left
Fig. 78(f) – The relationship between the external lamina (EL) and amorphous connective tissue material at the base of *P. lamarkii* epidermal cells. Microfilament bundles (f) terminate in hemidesmosomes (HD) and there is a close association of the connective tissue material opposite the hemidesmosomes (arrows). × 32 000. (Dr A. Bubel.)
SF, striated fibrils.

Above left
Fig. 78(g) – The close association and alignment of connective tissue striated fibrils (SF), which possess an ill-defined axial periodicity (arrows), along the plasma membrane (large arrows) of *P. lamarkii* epidermal cells. × 71 000. (Dr A. Bubel.)

Above
Fig. 78(h) – Thick connective tissue fibrils of the connective tissue of *P. lamarkii*, which possess a more distinct axial periodicity (arrows). × 95 000. (Dr A. Bubel.)

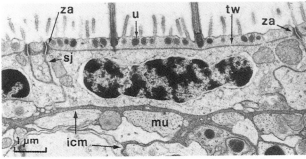

Fig. 79(c) – The cuboidal epithelium of the acoel *Haplopharynx* sp., which possesses a well-developed terminal web (tw). × 8 000. (Dr S. Tyler.)
icm, intercellular matrix.

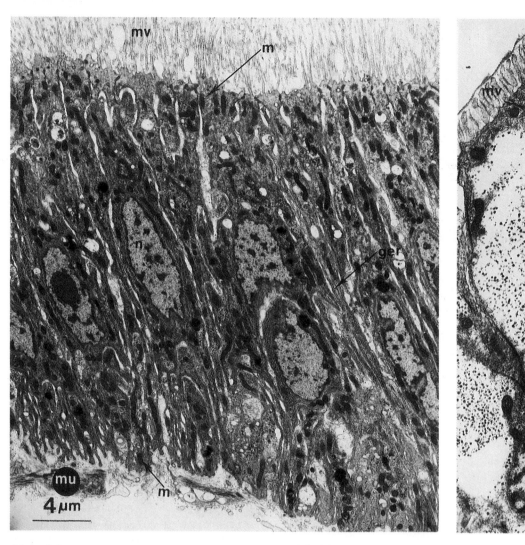

mv
m
n
ger
gl
mu
m
4 µm

Above left
Fig. 80(a) – The columnar epithelium of the outer mantle fold of the bivalve *Mytilus edulis*. × 4 000. (Dr A. Bubel.)

Above right
Fig. 80(b) – The low columnar epithelium of the outer mantle fold of the bivalve *Cardium edule*. × 10 000. (Dr A. Bubel.)

such cells is occupied by microtubules that terminate in hemidesmosomes (Figs 83). It is, however, not clear whether the microtubules transmit tension from the muscle to the exoskeleton passively, via a cytoskeletal role, or are more actively involved in the modulating of the tension via movement.

BASEMENT MEMBRANE

The basement membrane is found immediately adjacent to the plasma membrane of most epithelial cells and encloses nerves, muscles and blood vessels and capillaries. It is continuous with the ground substance of the connective tissue, also. A continuous sheet of basement membrane is thought to play an important structural role. It is considered to function as a filter in some situations, e.g. in annelid blood vessels and insect Malpighian tubule cells, where it is strategically positioned to regulate the ionic and molecular traffic between tissues and blood. In arthropods, for example, the epidermis is separated from the haemocoel by a basement membrane (basal lamina), which is a charged molecular sieve

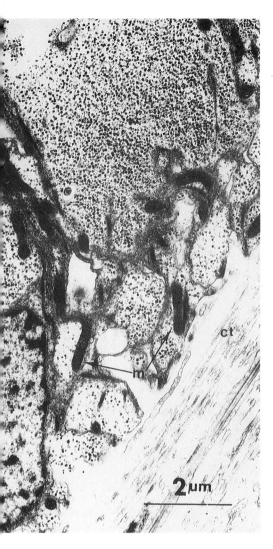

ct

m

2 um

and invertebrates is between 20 and 300 nm thick. The lamina reticularis (synonyms: recticular lamina; microfibrillar layer) appears to contain reticular fibres, microfibrils and fine filamentous elements (Fig 77). It may be quite thick in turbellarian platyhelminthes, between 10 and 14 μm, and may show a highly organized helical pattern of fibre packing.

A luminal basal lamina, or vascular lamina, composed of extracellular material between 20 and 300 nm thick lines the lumina of invertebrate blood vessels, e.g. in brachiopods and annelids (Figs 142a, 144a). The vascular lamina in the annelid, *Amythas hawayanus* is composed of (a) a lamina densa, composed of thin sheets of electron-dense material, (b) a lamina rara, and (c) an extensive reticular lamina, consisting of cross-banded collagen fibres embedded in a matrix.

It appears that the basement membrane mediates the connection between the plasma membrane of epithelial cells and the extracellular connective tissue elements. The formation of cross-bridges in the lamina lucida apparently connects the epithelial cell plasma membrane and the basal lamina. In the rat capillary endothelial basement membrane, short cross-bridge filaments, 10–15 nm in diameter and from 30 to 90 nm in length, are distributed at intervals of 40–60 nm in the lamina lucida to connect the plasma membrane and basal lamina. In the lamina reticularis, micro-fibrils which are anchored in the basal lamina are connected to collagen fibrils of the connective tissue (Fig 77). In contrast, in the epithelial basement membrane of the salivary gland, filaments of 17 nm thickness and 40 nm length traverse the lamina lucida and are connected to globular particles about 30 nm in diameter in the basal lamina and to dense plaques on the plasma membrane. Filamentous/amorphous materials also connect electron-dense plaques/hemidesmosomes on the plasma membrane of hair-follicle sheath cells and insect Malpighian tubule cells to their basal laminae (Figs 12, 106b).

It is thought that the basal lamina is composed of a fine fibrillar collagenous component and an amorphous non-collagenous material. Biochemical evidence suggests the existence of laminin, a non-collagenous glycoprotein, in the lamina lucida. It is assumed that laminin is an attachment protein between the plasma membrane and the basal lamina. Also, in the

having pores comparable in size to large haemolymph proteins.

The basement membrane has a three-layered structure made up of (a) the lamina lucida, (b) the lamina basalis or basal lamina, and (c) the lamina reticularis. The lamina lucida, which faces the plasma membrane, usually appears electron-transparent, but with tannic acid or ruthenium staining appears fuzzy and filamentous. The basal lamina (synonyms: limiting membrane; boundary membrane; external lamina; lamina densa) is structurally complex, being composed of a matrix within which granules, fibres and fine filaments are embedded. In some situations, e.g. rat capillary endothelium, it often appears double-layered and fine cross-linker filaments span the lamina. The basal lamina in both vertebrates lamina lucida, the coexistence of laminin and collagen type IV has been demonstrated.

CONNECTIVE TISSUE

Cells rest on a basement membrane, which consists of a basal lamina and filamentous and fibrous components embedded in an amorphous ground material. The filamentous material is scattered throughout the ground substance without any apparent orientation (Fig. 78d). However, the fibrous component is composed of fibrils which range from 10 to 100 nm in diameter and may show, depending on their location, some organization into orthogonal layers (Fig. 78e) or into two layers, an outer longitudinal layer and an inner circumferential layer (Fig 78a–c). The connective tissue fibrils of invertebrates in cross-section possess a dark periphery with a lighter or central core, and may also possess an axial periodicity. The axial periodicities of these fibrils are considerably shorter than the usual 64 nm vertebrate pattern. They vary considerably; those noted are 7 nm, 10–15 nm, 30 nm, 50 nm, 56 nm, 28–60 nm, and 64 nm (Fig 78b–h). It is not clear whether the variations observed in invertebrate connective tissues are an indication of differences inherent in the collagen of different invertebrates, or whether they are simply an expression of shrinkage caused by fixation, embedding and sectioning techniques employed during their preparation for electron microscopy.

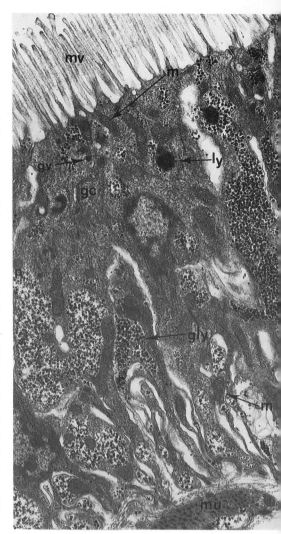

EPITHELIAL CELLS

Epithelial tissues compose the surface and lining of the animal body. They form a continuous layer or sheet over the entire body surface and most inner cavities. On the external surface, epithelial tissues protect underlying cells from injury, bacteria, harmful chemicals and drying. On the internal surface, epithelial tissues absorb water, nutrients and give off waste products. A basement membrane forms a support for an epithelium and connects it to an underlying connective tissue.

Epithelial tissues can be divided into three major groups according to the shape of the cells composing them:

(1) Squamous epithelium is composed of thin, flat cells with only a centrally situated nucleus causing a slight bulge towards the free surface, e.g. lung endothelium (Figs 108, 109).

(2) Cubiodal epithelium is made up of cube-shaped cells with round centrally placed nuclei and usually, at the free border, short microvilli, e.g. kidney cells (Figs 81b, 103b).

(3) Columnar epithelium cells resemble columns with nuclei usually situated near the bases of the cells. These cells are widespread (Figs 80, 81a). Scattered throughout columnar epithelial cells are glandular cells, specialized to secrete substances such as mucus (Figs 98, 99).

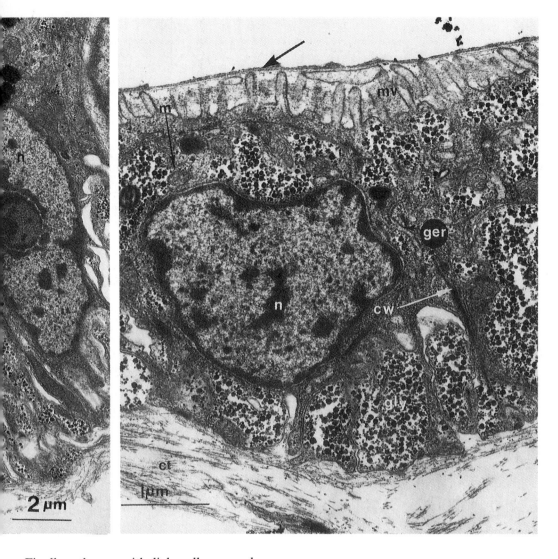

Above left
Fig. 81(a) – The low columnar epithelium of the central zone of the mantle of *C. edule* with the cell cytoplasm packed with glycogen (gly). × 7 500. (Dr A. Bubel.)

Above right
Fig. 81(b) – The cuboidal epithelium of the pallial mantle region of the bivalve *C. edule*. Extrapallial material is associated with the microvilli (mv) (arrow). × 24 000. (Dr A. Bubel.)

Finally, these epithelial cells may be specialized into a sensory epithelium to receive stimuli or have single sensory cells scattered throughout.

A basement membrane usually forms a support for an epithelium. However epithelia in some invertebrates lack a basement membrane (Fig 79). In such epithelia there is a tendency for the nuclei to lie at a level even with or below body wall muscles. Epithelial cells such as these are referred to as insunk or infranucleate cells (Fig 82). Generally, epithelial cells are geometrically regular, having the shape of flat polygons, cubes or polygonal cylinders. They are held together by the stickiness characteristic of all cells, and by cohesive forces between the parallel surfaces of

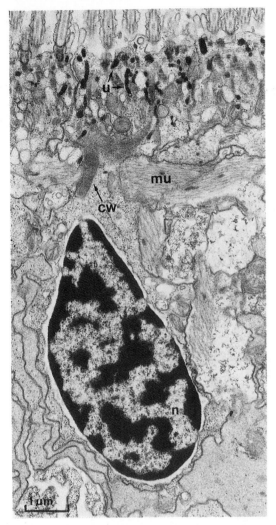

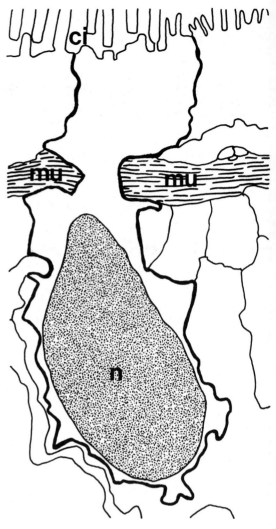

adjacent cells. In addition, between adjacent cells there are regions of lateral membranes which are specialised to give firm attachment (Figs 3, 4, 5, 6, 7, 8). There are many epithelial variations in the elaboration of intracellular cytoskeletal structures, of extracellular matrices and of secretory products (Figs 83, 84). The free surface of epithelial cells are, also, modified according to the local functional demands.

TENDON CELLS

Myoepithelial junctions in invertebrates occur at diverse loci to provide the necessary connections for muscular activity. The muscles of invertebrates are attached to exoskeletal structures by way of specialized epithelial cells

Above left
Fig. 82(a) – An insunk epithelial cell of the acoel *Paratomella sp.*, bearing adhesive cilia and a well-developed cell web (cw). × 10 500. (Dr S. Tyler.)

Above
Fig. 82(b) – Diagram of Fig. 82(a).

Right
Fig. 83(a) – A tendon cell, underlying the mural plate of the barnacle, *Elminius modestus*, containing tightly packed microtubules (mt) and exoskeletal rods (exr) in the apical region, with a much folded basal membrane lined with hemidesmosomes (hd). × 45 000. Inset: free ending microtubules in the apical region of a tendon cell. × 100 000. (Dr A. Bubel.)
chd, conical hemidesmosome.

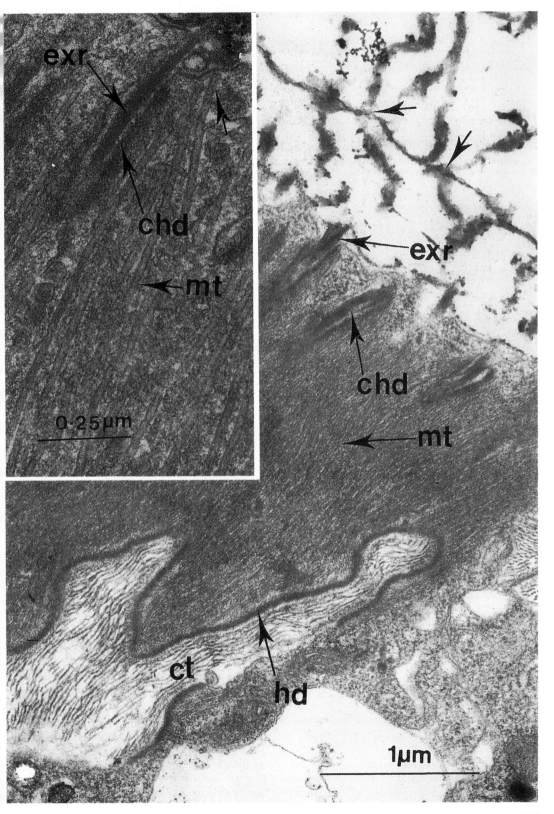

Below
Fig. 83(b) – The basal region of a tendon cell of *E. modestus*. The folded basal membrane is lined with hemidesmosomes on which microtubules (mt) terminate. × 58 000. (Dr A. Bubel.)

Right
Fig. 84(a) – The relationship between tendon cells (TD), a cuticular flange (CFL) and peduncle muscles (PM) in the polychaete, *Pomatoceros lamarkii*. Between the tendon cell bases and the terminal ends of peduncle muscles lined by hemidesmosomes (hd) is a myo-tendon zone (MTZ) composed of collagen fibrils. × 12 000. (Dr A. Bubel.)

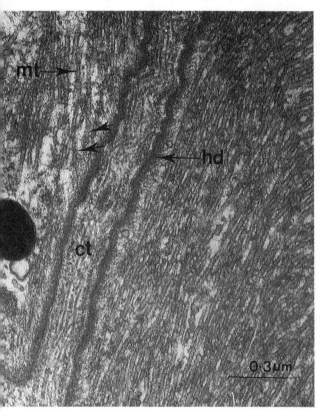

which are frequently referred to as 'tendon cells'. There are, however, differences in the tendon cells and the manner of muscle attachment in invertebrates. The tendon cells in most arthropods contain numerous microtubules which articulate with the exoskeleton by means of extracellular 'muscle attachment fibres'. The great bulk of the cytoplasm of such cells is occupied by microtubules; other organelles are scarce. The microtubules (each 23 nm in diameter), extend from the apical to the basal regions of the cells and terminate in hemidesmosomes (Fig 83a, b). The apical plasma membrane of these cells forms a series of invaginations, which are roughly cylindrical. Lining the adepithelial side of the invaginations is a conical hemidesmosome (Fig 83a). Electron-dense rods, which originate in these invaginations, pass into the exoskeletal structure (Fig 83a). Only a single rod extends from each conical hemidesmosome. In contrast, in molluscs and annelids, tendon cells are characterized by bundles of filaments and are attached to the shell and cuticle by a tendon sheath and microvilli, respectively (Fig 84a, b). At both apical and basal regions of the cells,

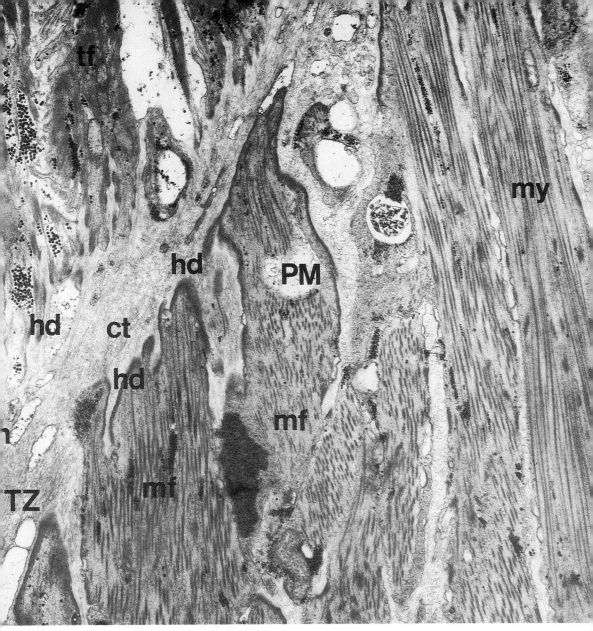

the filaments are inserted into hemidesmosomes (Fig 85). Muscles in arthropods and in some annelids are attached directly to the bases of tendon cells in highly differentiated junctional areas, whereas in molluscs and some other annelids, tendon cell bases and the terminal ends of muscles terminate independently of each other by means of hemidesmosomes in a myotendon zone filled with collagen fibrils (Fig 85). Although the absolute strengths of invertebrate muscle/skeletal attachments are not known, there is the possibility that, since some tendon cells contain

microtubules while others do not, there may be a direct relationship between the muscle tension exerted on a tendon cell and the presence of microtubules within the cells.

REFLECTOR CELLS

In the epipelagic and littoral species of cephalopods chromatic elements are located in the dermis. These include layers of physiologically alterable chromatophore organs and subjacent layers of reflector cells or iridophores (Fig

133

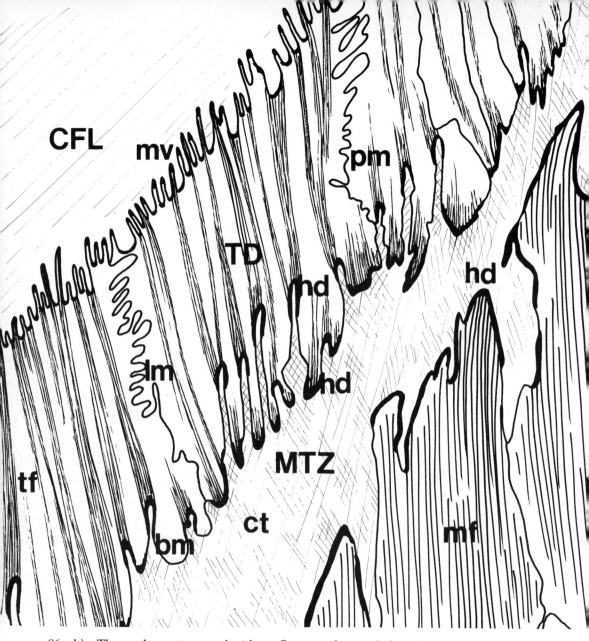

CFL   mv   pm   TD   hd   hd   lm   hd   MTZ   tf   ct   bm   mf

86a–b). These elements can absorb, reflect, diffract and scatter light. Dermal iridophores are multilayered structures based on plates of quarter-wavelength reflectors that act as broad-band reflectors. In cephalopods, iridophores are also important around the eye, the ink sac and the photophores. In the octopus, *Octopus dolfeini*, there are dermal reflector cells in the same position as iridophores in the dermis of squids and cuttlefishes. These are flattened cells which are orientated parallel to the surface of the skin. The functional optical units in the reflector cells and in iridophores of other cephalopods are known as reflectosomes and iridosomes, respectively. Both are composed of arrays of thin platelets, with a high refractive index, that are uniformly separated from one another by a gap of low refractive index. The reflector cells of octopi contain numerous reflectosomes, composed of discoidal lamellae orientated with their flattened surfaces in register like a stack of coins, which radiate out from the cell body to form a circumferential field (Fig 87a, b). Each lamellae consists of a discrete cytoplasmic inclusion, the reflecting platelet, within a fold

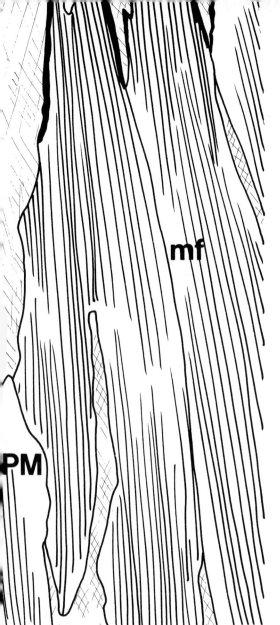

mf

PM

Left
Fig. 84(b) – Diagram of Fig. 84(a).

Below
Fig. 85 – Peduncle muscle fibres terminating in the myo-tendon zone of *P. lamarkii.* × 45 000. (Dr A. Bubel.) col, collagen fibrils; el, external lamina; tm, thick myofilaments.

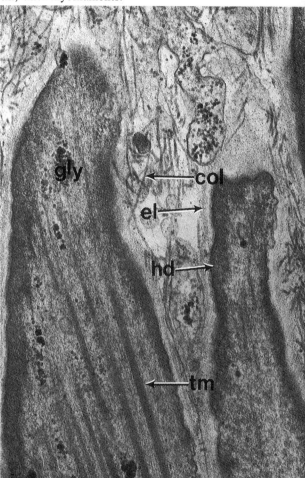

gly

col

el

hd

tm

of the plasmalemma. The plasmalemma is so closely apposed to the surface of the reflecting platelets that all cytoplasm is excluded, except in the neck region, where the reflecting platelet is attached to the cell body (Fig 87c). A reflecting lamella is, in effect, a tripartite structure composed of a continuous covering of plasmalemma, an incipient cytoplasmic space, and an inner reflecting platelet.

Each reflectosome is envisaged as a stack of discoidal thin films of alternating high and low refractive index. The band width and character of light reflected from each such unit depends on the refractive index, the thickness of each layer, the angle of illumination and the number of platelets in a stack.

SENSORY CELLS

Sensory cells are usually found in epithelia, scattered singly, aggregated in small groups, or even organized into tissues or organs which are recognized as sensory. Sensory cells in invertebrates are mostly primary receptor cells; bipolar nerve cells with cell bodies situated

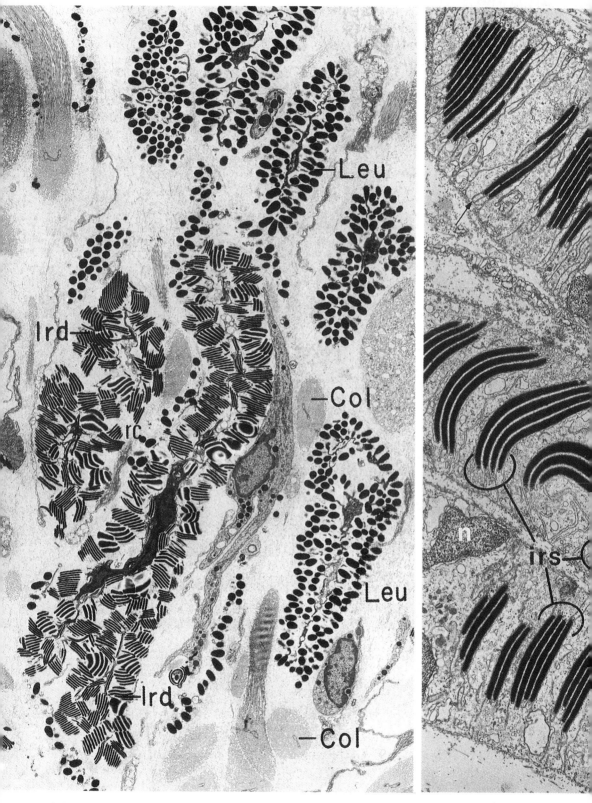

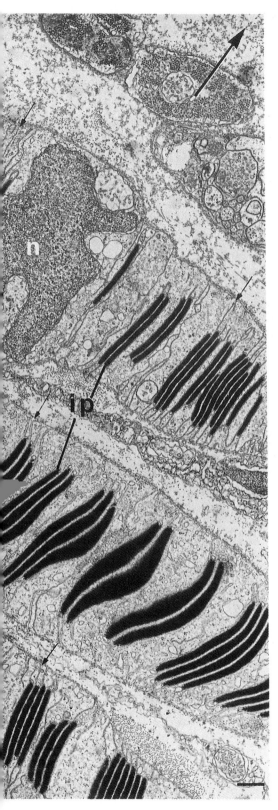

either below or in an epidermis. From each cell body a single dendrite extends distally to the body surface and usually bears microvilli and cilia, whilst proximally a single axon leaves its base (Fig 88a–d). These are mostly columnar cells that bear one or several cilia, which arise from basal bodies associated with ciliary rootlets. Some sensory cells lack cilia, although centrioles and ciliary rootlets may be present in their apical cytoplasm. The cilium of a typical sensory cell possesses a 9+2 axonemal pattern; however, cilia of some sensory cells lack some of the central or peripheral microtubular fibrils. Generally, sensory cilia are non-motile and their basal feet show no preferred orientation. A structural indication of non-motility in sensory cilia is thought to be the possession of microtubular fibrils, which terminate on the ciliary membrane. The cilia are usually between 3 and 5 μm long, but, in some cells, 10 μm or more in length. Sensory cells are connected to supporting cells by zonulae adhaerentes and septate junctions, by septate junctions alone, or by zonula specializations;

Far left
Fig. 86(a) – Reflector cells and leukophores in the dorsal tegument of the mollusc, *Octopus dolfeini*. Both reflector cells (rc) and leukophores (Leu) have central cell bodies and a complex array of peripheral cytoplasmic inclusions that are joined to the cell bodies by narrow stalks. In the reflector cells the discoidal reflecting lamellae are organized into stacks called reflectosomes that reflect blue and green light. Ovoid inclusions in the leukophores scatter all wavelengths of light. Bundles of collagen fibrils (Col) are found throughout the dermis. × 4 800. (Dr S. Brocco, University of Washington.)
Ird, iridophores.

Left
Fig. 86(b) – Iridophores of the mollusc *Loligo opalescens*. Parts of three iridophores are sectioned transversely. The large arrow indicates the direction of the epidermis. Groups of iridosomal platelets (ip) are arranged in sets called iridosomes (irs). The platelets seen in transverse section are really long tapered ribbon-shaped structures. Each platelet is intracellular but lies next to folds of the plasma membrane that borders it on both lateral surfaces. Infoldings of the plasma membrane are indicated in several places by small arrows. The apparently thick platelets in the central cell body are oblique sections through long ribbons that almost always follow a serpentine pathway along the length of the iridophore. × 7 700. (Dr S. Brocco.)

137

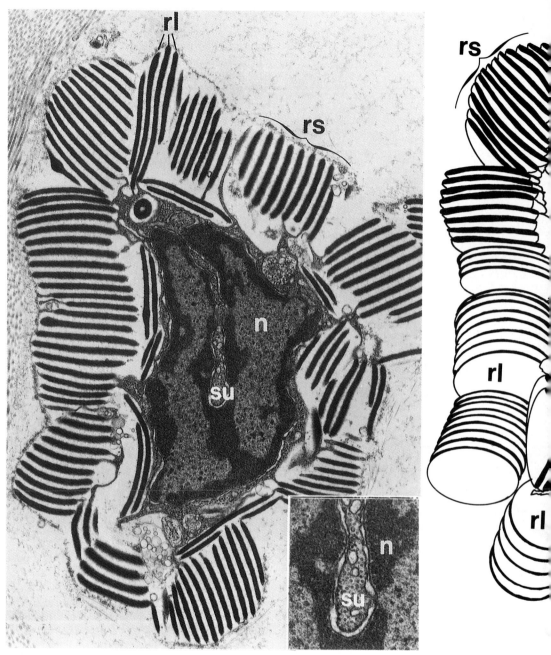

Left

Fig. 87(a) – Transverse section through a reflector cell. The reflectosomes (rs) radiate from the cell body of a reflector cell. Each reflectosome consists of a file of discoidal reflecting lamellae (rl) parallel to and in register with one another. Adjacent reflecting lamellae are separated from one another by an inter-lamellar space that is confluent with the extracellular connective tissue compartment. The nucleus (n) has a deep suncus (su) along its length. × 16 000. (Dr S. Brocco.) Inset: Nuclear sulcus. × 34 000.

Centre

Fig. 87(b) – Diagram of 87(a).

Right

Fig. 87(c) – Section of a developing reflector cell. Each platelet is connected with the cytoplasm by a stalk (ringed). × 19 300. (Dr S. Brocco.)

GB, Golgi body; np, nuclear pore; Nu, nucleus; V, vesicles.

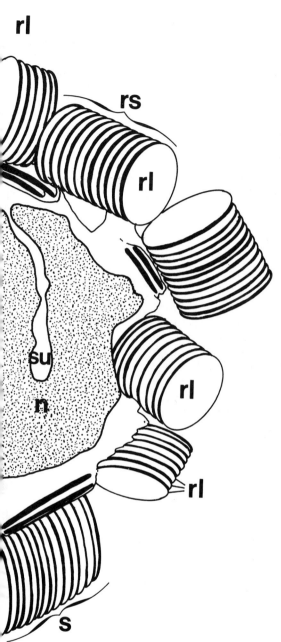

rl

rs

rl

rl

rl

su

n

rl

s

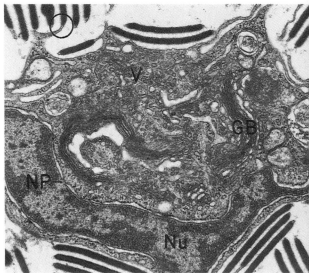

V

GB

NP

Nu

zonula adhaerens, and zonula intermedia. Some receptor cells are connected by tight and gap junctions to accessory cells resembling epidermal cells. Their cytoplasm contains a nucleus, Golgi apparatus, many vesicles, some or no granular ER, free ribosomes, a little agranular ER, microtubules and a large number of mitochondria. Within the axonal process of sensory cells, the cytoplasm includes microtubules, vesicles and mitochondria. For each sensory cell, the diagnostic feature is a basally produced axon.

Sensory cells in arthropods, unlike those in other invertebrates, are found in subcuticular receptor organs called sensilla. These appear in the form of hairs (setae), pit-pegs or plates on the exoskeleton. Usually, a small number of primary receptor cells (less than 10) and three auxiliary cells (the tormogen, trichogen and thecogen) are combined to form a sensilla. The receptor cells are bipolar nerve cells with distal dendrites that provide the sites of stimulus transduction and proximal axons. A modified ciliary structure, the ciliary segment, separates an inner dendritic segment (which contains most cell organelles) from an outer dendritic segment (which contains microtubules, only) Figs 89a–c, 90a–d). The ciliary segment has a 9+0 axonemal pattern and possesses a single basal body from which a ciliary rootlet extends into the inner dendrite segment. Microtubular fibrils, which may have electron-dense cores, extend for a shorter distance into the outer dendrite segment. The outer dendrite segment is bathed in an extracellular fluid, which is secreted by the non-neural auxiliary cells (Fig 89a–c). The auxiliary cells are also responsible for the formation of the specialized cuticle covering the receptor endings.

It is rarely possible to establish the sensory nature of the free nerve ending by tracing its dendrite to the cell body. There is always the danger of confusing a specialized non-sensory cell of 'sensory appearance' with a sensory cell. It is often difficult to distinguish between

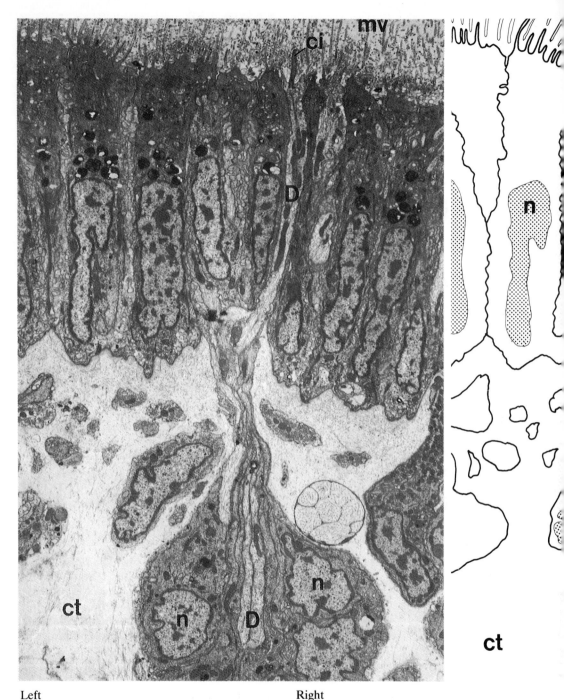

Left
Fig. 88(a) – A dendrite (D) extending from a ganglion and terminating as a type 1 ciliated ending (ci) at
the surface of the lip of the mollusc, *Aplysia
brasiliana*. × 7 000. (Dr D.G. Emery, University of
Texas.)

Right
Fig. 88(c) – Type 2 sensory ending in the lip of *A.
brasiliana*. The dendritic process (D) is densely
packed with microtubules and has several mitochondria (m) just below the apical tuft of cilia (ci). ×
10 500. (Dr D.G. Emery.)

Centre
Fig. 88(b) – Diagram of Fig. 88(a).

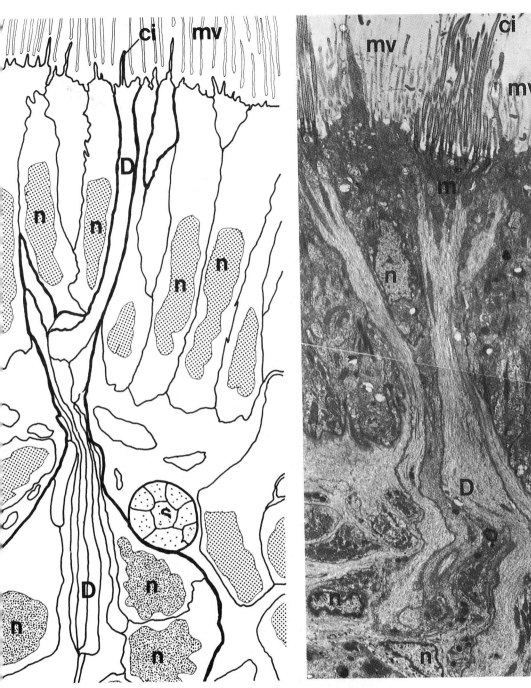

mechanoreceptors, chemoreceptors and photoreceptors solely on the basis of structure. There is no reliable ultrastructural criterion whereby the modality of these kinds of receptor can be determined. An interpretation of the function of different receptor types is also obscured by the morphological variation that exists within each type.

Mechanoreceptors

Mechanoreceptor cells are generally ciliated sensory cells. The number of cilia per cell and details of their ultrastructure show some variation. The number of cilia borne by the cells ranges from 1 to 5–8, to 10–40 and some possess a 9+2 axonemal structure with basal bodies, basal feet and striated rootlets,

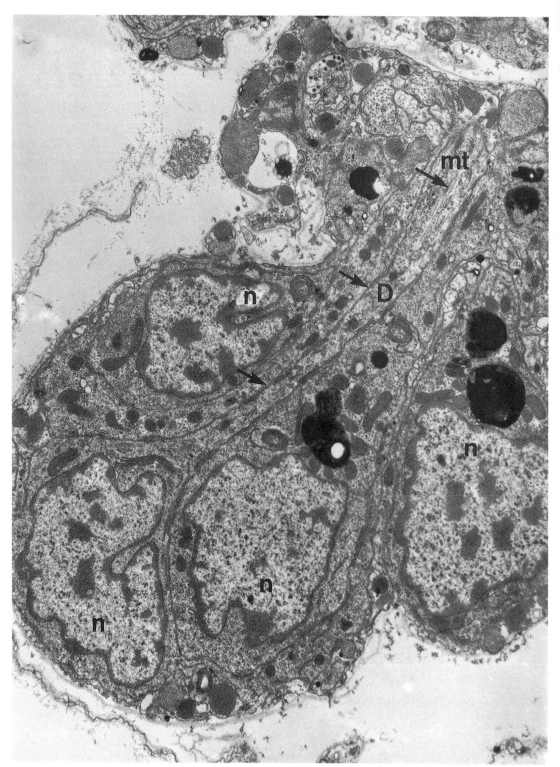

Fig. 88(d) – Sensory ganglion in the lip of *A. brasiliana*. The dendrite extends from a neuron at the base of the ganglion. Microtubules (arrow) begin to appear in the dendritic cytoplasm near the distal pole of the ganglion. × 11 500. (Dr D.G. Emery.)

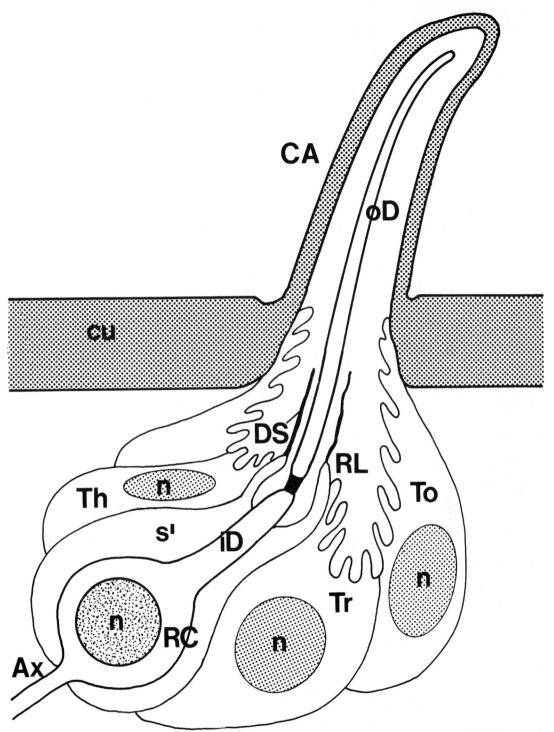

Fig. 89(a) – Diagrammatic representation of an insect chemo- , hygro- , thermoreceptor sensilla, in longitudinal section. Only one receptor cell (RC) is shown with the axon (Ax) and the dendrite, which is subdivided
into an inner (iD) and outer (oD) segment by a short ciliary segment (black). Three non-neural auxilliary
cells surround the receptor cell (s^1) and border the receptor lymph cavities (RL). The thecogen (Th) forms
the dendrite sheath (DS) and wraps the inner dendritic segment and receptor cell soma, the trichogen (Tr)
and tormogen cell (To) form the cuticular apparatus (CA) during ontogeny. (After Dr R.A. Steinbrecht.)

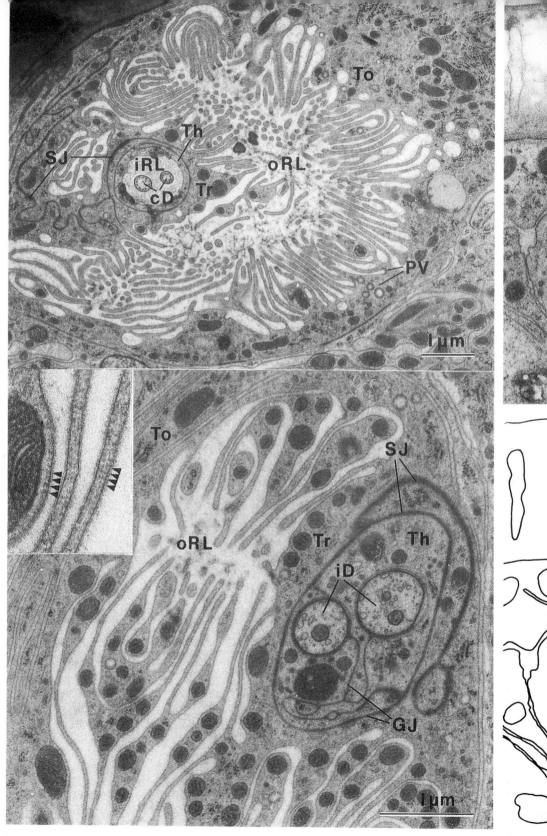

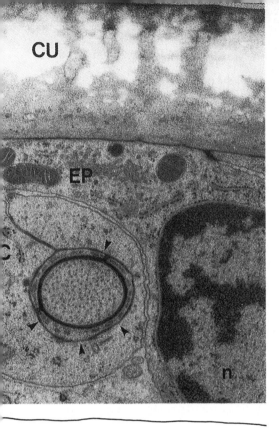

CU

EP.

n

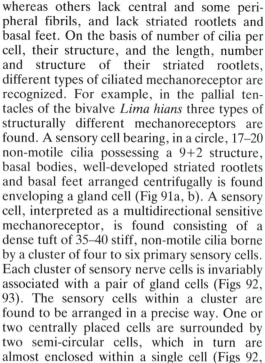

whereas others lack central and some peripheral fibrils, and lack striated rootlets and basal feet. On the basis of number of cilia per cell, their structure, and the length, number and structure of their striated rootlets, different types of ciliated mechanoreceptor are recognized. For example, in the pallial tentacles of the bivalve *Lima hians* three types of structurally different mechanoreceptors are found. A sensory cell bearing, in a circle, 17–20 non-motile cilia possessing a 9+2 structure, basal bodies, well-developed striated rootlets and basal feet arranged centrifugally is found enveloping a gland cell (Fig 91a, b). A sensory cell, interpreted as a multidirectional sensitive mechanoreceptor, is found consisting of a dense tuft of 35–40 stiff, non-motile cilia borne by a cluster of four to six primary sensory cells. Each cluster of sensory nerve cells is invariably associated with a pair of gland cells (Figs 92, 93). The sensory cells within a cluster are found to be arranged in a precise way. One or two centrally placed cells are surrounded by two semi-circular cells, which in turn are almost enclosed within a single cell (Figs 92,

Above far left
Fig. 89(b) – A cross-section of a trichodeum of the silk moth *Bombyx mori* at the level of the ciliary segments. Two ciliary segments (cD) are found in a separate inner receptor lymph cavity (iRL) formed by the thecogen cell (Th). The outer receptor lymph cavity (oRL) at this level is mainly bordered by the tormogen cell (To) which displays extensive microvilli and microlamellae. × 15 000. (Dr R.A. Steinbrecht, Max-Planck-Institute.)

Below far left
Fig. 89(c) – The inner dendritic segments (iD) of *B. mori* trichodeum are wrapped by the thecogen cell (Th). The large outer receptor lymph cavity (oRL) is bordered by extensive microlamellae of the trichogen cell (Tr), which at their cytoplasmic side are studded with portasomes (arrowheads on inset). × 26 000. (Dr R.A. Steinbrecht.)

Above left
Fig. 90(a) – The concentric wrapping of the dendritic outer segment, of the subcuticular receptor of the isopod *Porcellio scaber* by inner (arrowheads) and outer enveloping cell (OEC). × 18 000. (Dr T. Haug, University of Regensburg.)
Cu, cuticle; E, epidermal cell.

Below left
Fig. 90(b) – Diagram of Fig. 90(a).

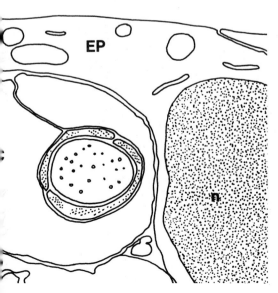

EP

n

145

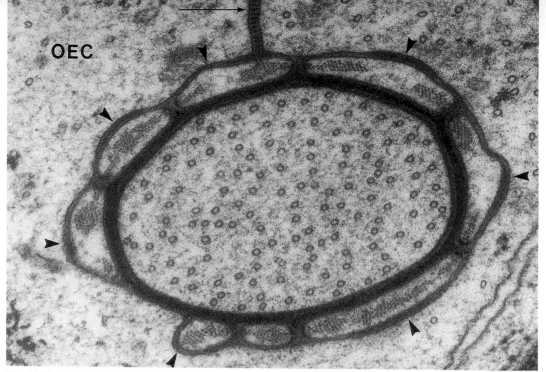

Fig. 90(c) – Transverse section through the dendritic outer segment of the subcuticular receptor of *P. scaber*, at the level of the apical projections of the inner enveloping cell (arrowheads). Bundles of filaments (5 nm diameter) are evident within the projections and a septate junction (arrow) is present at the mesaxonic closing of the outer enveloping cell (OEC). × 114 000. (Dr T. Haug.)

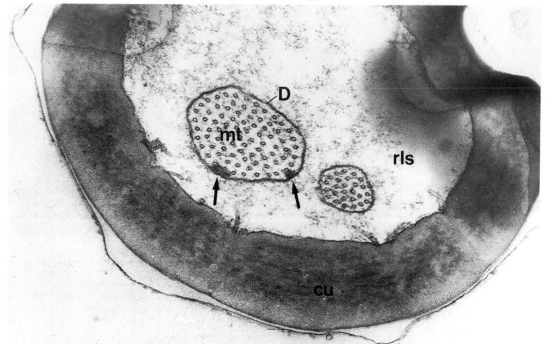

Fig. 90(d) – Cross-section of an olfactory hair from the silk moth *Antherea polyphemus*. Electron-dense structures (arrows) are found between the membrane and microtubules (mt) in a large dendrite (D). × 45 000. (Dr T.S. Keil, Max-Planck-Institute.)
cu, hair cuticle; rls, receptor lymph space.

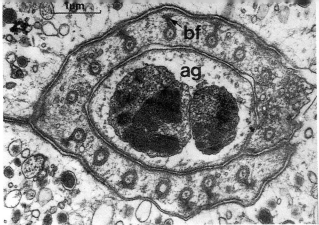

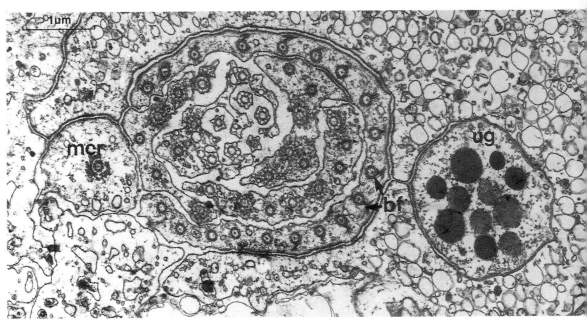

Above left
Fig. 91(a) – A section through the distal region of type A sensory complex near the apex of a cell in the pallial tentacles of bivalve, *Lima hians*. The cilia arise from a single sensory cell which encircles a gland cell (ag). × 14 000. (Prof G. Owen.)
bf, basal feet.

Above right
Fig. 91(b) – Type A sensory tuft: the area within the circle of cilia represents the opening of a gland cell. × 12 000. (Prof. G. Owen.)

Below
Fig. 92 – A section in a plane parallel to the surface of a pallial tentacle of *L. hians* through a type B sensory tuft, revealing the orientation of the basal feet (bf) and the positions of the monociliary sensory cell (mcr) and unpaired gland cell (ug). × 20 000. (Prof. G. Owen.)

93). The cilia are 5 μm in length, possess the normal 9+2 fibril complement and well-developed rootlets. However, the point of particular interest is the orientation of the central singlets and of the basal feet, one of which projects laterally from the basal body of each cilium. The basal foot of each cilium tends to lie in a plane which radiates from the centre of the cluster of ciliated cells (Fig 93). It is suggested that the sensory cells are depolarized when the sensory cilia are displaced in the direction of the basal foot. The basal feet in these cilia lie at right angles to the plane joining the two central singlets. In motile cilia, the effective or power stroke lies in this plane. Thus, displacement of such a ciliated tuft in any direction will excite at least some cilia comprising the tuft. A sensory cell, believed to

147

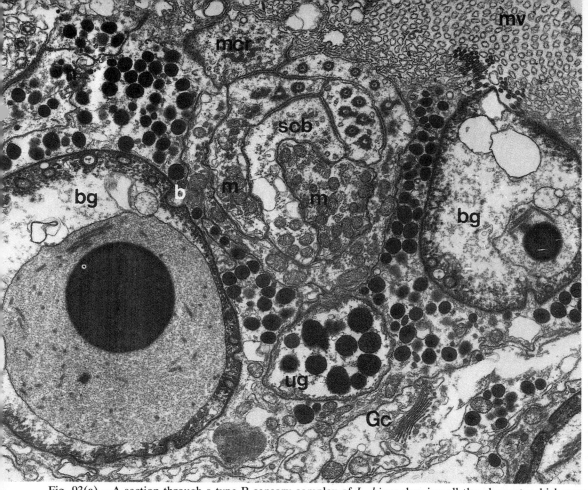

Fig. 93(a) – A section through a type B sensory complex of *L. hians* showing all the elements which may comprise a type B sensory complex. These are: (1) the cluster of sensory neurons (scb), which bear non-motile cilia; (2) a single sensory neuron (mcr) which bears the kinocilium and nine stereocilia; (3) paired gland cells (bg) and (4) an unpaired gland cell (ug). × 14 000. (Prof. G. Owen.)

be a vibration receptor, is frequently found close to the tuft of sensory cilia. This receptor bears a single stiff long cilium(10–15 μm), or kinocilium, surrounded by shorter modified microvilli, or stereocilia (Fig 94a). This receptor has a complex basal structure. Well-developed rootlets arise, not from the basal body, but from a short cylinder of fibrous material that encircles the basal body. The basal body possesses the typical arrangement of triplet fibrils and gives rise to nine radiating spokes, which extend to the fibrous cylinder and to alternate stereocilia. Vibration receptors with similar complex basal structures are also found in ctenophora. Other mono-ciliated sensory cells believed to be mechano-receptors are found in *Priapulus caudatus*. In this organism, the cilia are characterized by stout branching rootlets linked to their basal

bodies and surrounded by seven stereocilia (Fig 94b).

Multiciliated sensory cells function as mechanoreceptors in protected internal sites, such as in statocysts. Statocysts of varying complexity are found in coelenterate medusae, annelids and molluscs. They are organs of balance and orientation. The basal feet of sensory cilia in some molluscan statocysts are orientated in a way that suggests that they are directionally sensitive. There is, however, an apparent absence of uniformly orientated basal feet on the cilia of most mechanoreceptors. This is to be expected, since it is doubtful whether the nervous system of soft-bodied invertebrates, can compute the relative positions of almost infinitely flexible parts. In annelids, sensory cells bearing 6–10 cilia with a 9+0 and 8+2 shaft structure, and possessing

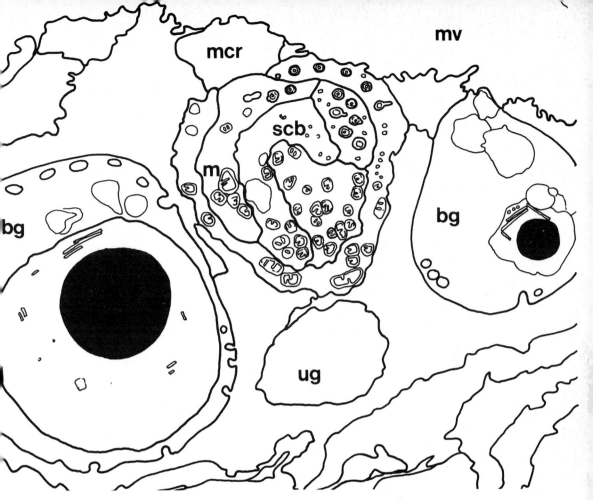

Fig. 93(b) – Diagram of Fig. 93(a).

basal bodies but no rootlets, are found lying parallel to the body surface below the cuticle. By virtue of the position of the cilia, a proprioreceptor function is attributed to the sensory cell.

Non-ciliated receptor cells of varying structure, which function as mechano-receptors, are found in the suckers of cephalopods. These receptors are involved in shape and/or negative-pressure discrimination.

Chemoreceptors

Both ciliated and non-ciliated chemoreceptor cells are found. Chemosensory cilia usually possess a 9+2 structure, basal bodies, and striated rootlets, but the possession of basal feet does not appear to be a universal feature. The number of cilia borne by chemosensory cells varies from 1 to 1–8, to more than 20 (Fig 88a–d). At least five morphologically different ciliated chemoreceptor cell types are found in the olfactory organs of cephalopods, namely: (1) those with slightly invaginated ciliated surfaces, (2) those with a distal cup-like ending (3) those with an internal cavity containing cilia, (4) those with a ciliated cavity that opens onto the body surface via a narrow neck, and (5) those with an enclosed ciliated cavity and possessing a distal process that has a ciliated ending at the epithelial body surface (Fig 95).

In some non-ciliated cells, microvilli are packed with tubular structures and, at their bases, single centriole-like structures are found.

Photoreceptors

Photoreceptor cells in invertebrates possess photoreceptive membranes of either ciliary or microvillar (rhabdomeric) origin. A variety of

149

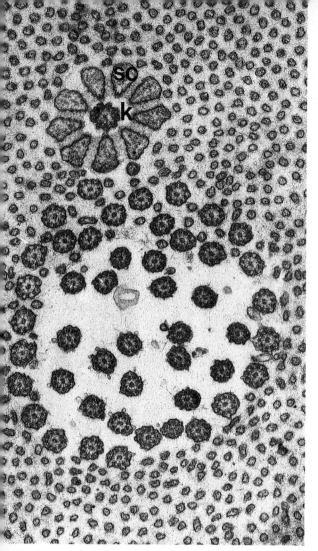

photoreceptors are found in invertebrates that range from single photoreceptor cells through simple ocelli to complex camera-type 'eyes'. Single-celled rhabdomeric photoreceptors may possess irregularly arranged long microvilli and independent cilia with a 9+0 structure and no rootlets. In contrast, ciliary photoreceptors may possess modified cilia with a 9+0 structure, basal feet and no rootlets, but which are flattened out to form a membrane array. In some photoreceptor cells cilia possess a 9+2 structure and their ciliary membranes develop extensive lateral expansions, which coil together to form so-called lamellate bodies. Two distinct photoreceptor types are found in some annelids, namely ocelli-composed receptor cells of the rhabdomeric type and brachial 'eyes' containing rhabdomeric and ciliary lamellate body photoreceptors (Fig 96a, b). In some insects there are internal ocelli composed of rhabdomeric photoreceptor cells which make contact with compound eyes (Fig 97a–c). The internal ocelli are believed to play a role as pacemakers of the circadian rhythm.

The morphologically different ciliary and rhabdomeric photoreceptor cells can be distinguished physiologically. Rhabdomeric photoreceptor cells, when illuminated, show a discharge of impulses and show an ON response, characteristic of visual organs. No impulses are discharged during illumination of ciliary receptors which show an OFF response. Discharges are only observed in ciliary receptors at cessation of illumination or reduction of its intensity. The OFF response is, however, strictly dependent for its excitation on the preceding period of illumination.

Above
Fig. 94(a) – A section through a type B sensory tuft and associated kinocilium (k) and stereocilium (sc) in a plane parallel to the surface of the pallial tentacles of *L. hians*. × 27 500. (Prof. G. Owen.)

Right
Fig. 94(b) – Cross-section through a receptor cell of the pseudocoelomate *Priapulus caudatus*. It is composed of a cilium, microvilli and surrounding sensory cell cytoplasm. × 10 000. (Dr V. Storch.)

Far right
Fig. 95 – Section through a large hollow receptor cell situated below the olfactory epithelium of the mollusc, *Octopus joubini* (juvenile). Basal bodies with feet (bb) are evident around the edge of the cilia(ci)-filled lumen. Several layers of glial-like supporting cell processes (glp) surround the re-receptor. × 9 800. (Dr G. Emery.)

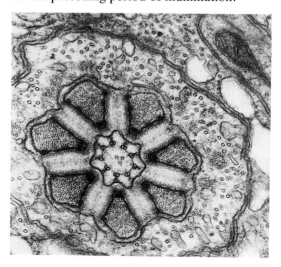

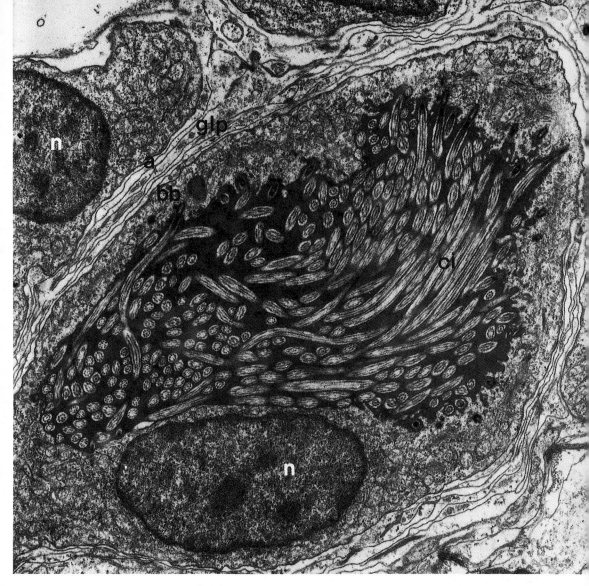

GLAND CELLS

Gland cells are found in many different locations in both vertebrates and invertebrates. These cells may be categorized into mucous and non-mucous types. Both cell types are characterized by numerous secretory granules of widely differing appearance (Figs 98a, b, 99a–d, 100). There is in most of these cells a well-developed granular ER and prominent Golgi complexes. In the pancreas two-thirds of the acinar epithelial cells are occupied by granular ER (Fig. 31a). The organelles appear to be involved in the synthesis of secretory granules. The granular ER and Golgi complexes show a topographical relationship which

is typical of secretory cells (Fig 101). The process of secretory product release varies according to cell type, e.g. by way of exocytosis or discharge of a huge intracellular mass of secretory product. Acinar epithelial cells of the pancreas discharge the contents of their zymogen granules (which contain enzymes or precursors of enzymes) by exocytosis, whereby the zymogen granules fuse with the luminal plasma membrane and the granules content is released. In other cells, microtubules arranged in a circle around the long necks of cells may play a role in the mechanism, by their contraction (Fig 75a). However, microtubules are associated with subsurface cisternae in the necks of some glands (Fig 75b). The function

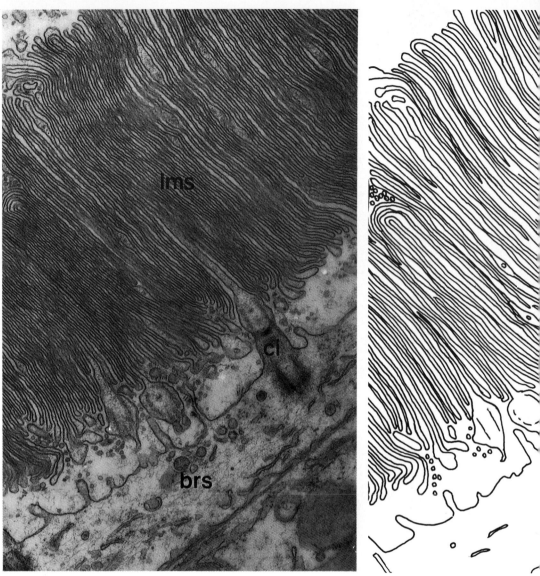

Left
Fig. 96(a) – The proximal sensory segment of the branchial receptor cell (brs) of the polychaete, *Spirobranchus giganteus* composed of a lamellar membrane stack (lms) produced by sensory cilia (ci). × 40 000. (Dr R.S. Smith, James Cook University.)

Centre
Fig. 96(b) – Diagram of Fig. 96(a).

Right
Fig. 97(a) – A longitudinal section through the internal ocellus I (IO) of the mothfly, *Psychoda cino*. There is different pigmentation of the compound eye (EY) and internal ocellus. × 3 000. (Dr P. Seifert, University of Munich.)

of this arrangement, although not known, is thought to be concerned with the determination of cell shape and the transport of secretory granules. There is some evidence in some cells of a coalescence of secretory granules at the cell pore (composed of a ring of microvilli), prior to extrusion (Fig 102a, b). In addition to the apical ring of pore microvilli, a distinct occlusor mechanism is present in some mucus cells. The sphincter comprises an inner sphincter ring of about 50 filaments (each 5 nm in diameter) which lies within the apical cytoplasm, just below the pore microvilli, and an outer ring lying within adjacent supporting cells (Fig. 102c). The arrangement is such that

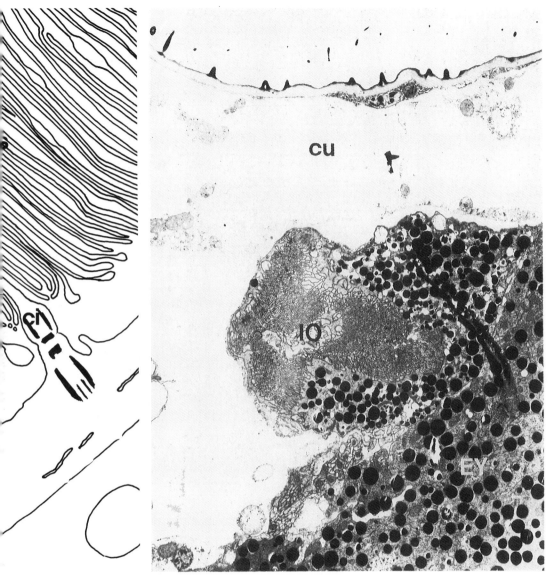

cu

IO

EY

the contraction of the ring filaments will bring about a closure of the pore, and their relaxation an opening of the pore.

TRANSPORTING CELLS

In vertebrates and invertebrates, excretory and transporting cells show many similar cyto-architectural characteristics, namely: greatly folded, apical, basal and lateral membrane surfaces, distended intercellular spaces, and a rich population of mitochondria. The chloride cell of the teleost gill and the distal segment kidney cells of the lamprey play important roles in ionic and osmotic regulation. Both cells are characterized by an intracellular branching tubular system and numerous mitochondria (Fig 103a, b). The luminal surface of the tubules of the lamprey kidney cells are covered by spirally wound parallel rows of particles (Fig 104a, b). These are believed to contain enzymatic activity which is essential for the osmoregulatory mechanism of the cell. In the teleost chloride cells, an increase in the rate of excretion by the cells in 200% sea water is accompanied by an increase in the tubular system (Fig 105a, b).

Similarly, in invertebrates, the Malpighian tubule cells of insects play an important role in

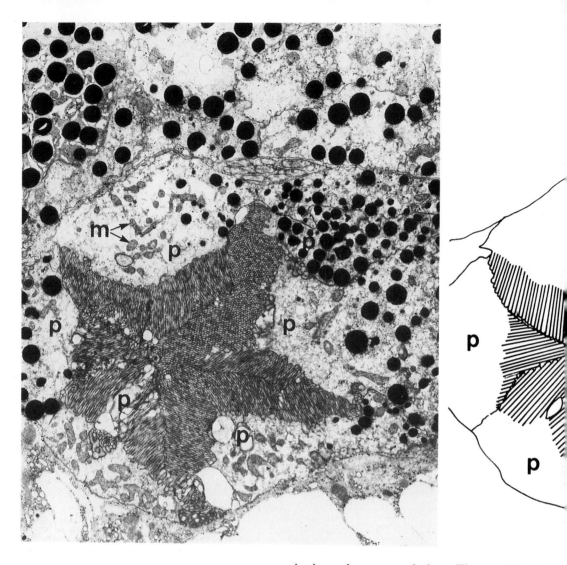

Above
Fig. 97(b) – Cross-section of internal ocellus 1 of the mothfly, *P. cino*. The rhabdomeres of six photoreceptors (p) form a star-shaped rhabdom in their apical regions. × 6 000. (Dr P. Seifert.)

Right
Fig. 97(c) – Diagram of Fig. 97(b).

ionic and osmoregulation. These cells show extremely amplified basal and apical surfaces and possess, in addition, a basal lamina which is strategically positioned to regulate ionic and molecular traffic between the tissues and blood (Fig 106a, b). In the pericardial gland cells of the mollusc, *Mytilus edulis* the basal cytoplasm is made up of numerous folds or cytoplasmic processes, which form a complex network of channels (Fig. 107a–c). These cells are involved in the uptake, degradation and removal of haemolymph constituents. The basal cytoplasmic processes terminate at a basal lamina and show a marked similarity to the basal region of podocytes from the mammalian glomerulus. These structural features may indicate that there is some selection in the takes place. Internally an alveolus is lined by a very thin squamous epithelium, whilst externally there is a dense network of blood capillaries embedded in a loose connective tissue. Adjacent alveoli share a capillary bed and connective-tissue space. A common wall, the alveolar septum, which may have local defects, alveolar pores, is composed of an inner alveolar lining, pulmonary capillaries and interstitial connective-tissue space (Fig 109a–b).

The inner alveolar lining is composed of squamous (type I cells; membraneous pneumocytes, respiratory cells), great alveolar (type II cells; granular pneumocytes; septal cells) cells, alveolar phagocytes and basal lamina. The squamous alveolar cells possess a large attenuated cytoplasm rich in ribosomes and micropinocytotic vesicles and few other organelles. In contrast, the great alveolar cells are cuboidal, with round large nuclei and a cytoplasm that contains numerous mitochondria, a few lysosomes, some granular ER and some specialized electron-dense bodies. These bodies are termed cytosomes or lamellar bodies. The great alveolar cells are considered to be secretory because cytosomes discharge their lipoprotein content onto the surface of squamous alveolar cells. The secreted material forms an essential part of the surfactant system in the lungs. The surfactant consists of 90% surface-active phospholipids, very rich in saturated lecithin, and 10% carbohydrate and protein. The surfactant reduces the surface tension of the alveoli, preventing collapse. Morphologically, this material may take the form of tubular myelin figures, derived from the phospholipid membranes of lamellar bodies. They are frequently found piled up as thick layers (Fig 110a–c).

Alveolar phagocytes (dust cells) move about on the surface of the squamous alveolar cells. Their structure is similar to that of wandering macrophages. The alveolar cells are seated on a thin, continuous basal lamina which creates a complete separation between the inner alveolar lining and the connective tissue interstitium of the alveolar system.

Pulmonary capillaries form a very dense network in the alveolar septum. The capillaries are lined by a squamous endothelium that is surrounded by its own basal lamina. The cytoplasm is thinly attenuated without fenestrations, but contains many micropinocytotic

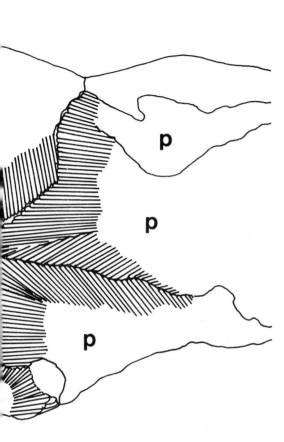

uptake and transport of materials. This selectivity may, in part, be determined both by a combination of the pores which occur between the basal cytoplasmic processes (or pedicels), and by selective filtration via the basal lamina.

RESPIRATORY TISSUE

In invertebrates, the exchange of gases may take place via the epidermis or via specialized respiratory organs: gills (Fig 108). However, in vertebrates, lungs represent the site of exchange of gases between air and the blood of pulmonary capillaries. Lungs are composed of millions of small alveoli, in which this exchange

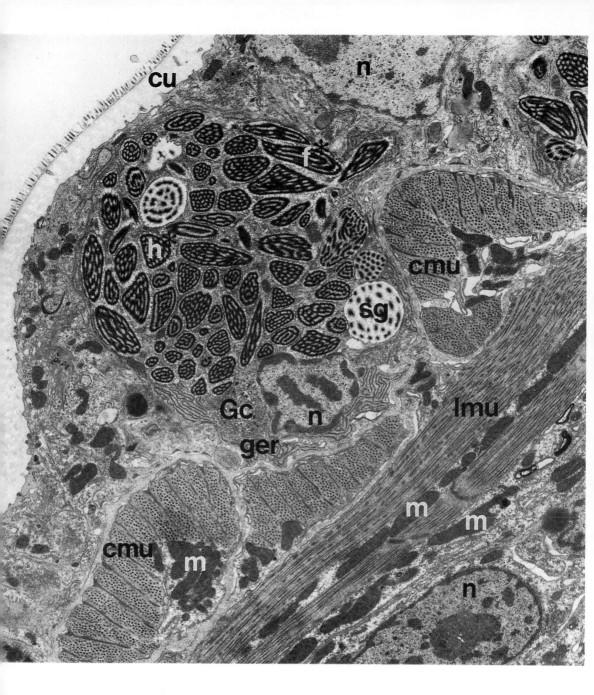

Above
Fig. 98(a) – Acid mucous cell of annelid, *Lumbricus vancouverensis* with secretory granules (sg) containing fingerprint (f*) and honeycomb-like contents (h*). The granules show varying degrees of condensation of secretory material. × 10 500. (Dr K.S. Richards, University of Keele.)
cmu, circular muscle; lmu, longitudinal muscle.

Right
Fig. 98(b) – Diagram of Fig. 98(a).

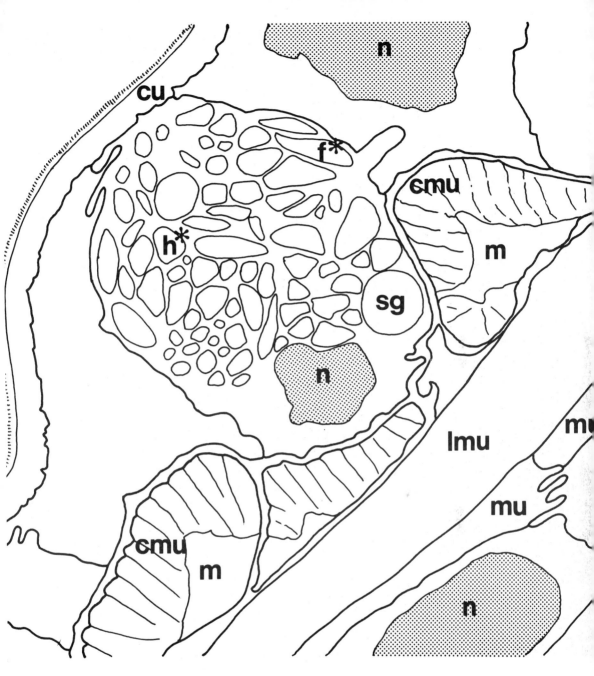

vesicles. The area of the endothelial cell facing the alveolus makes close contact with the squamous alveolar cell, and the basal laminae of the alveolar cell and endothelium often fuse. This area of close contact represents the blood–air barrier (Fig 109a, b).

The interstitial connective-tissue space of the alveolar septum is made up of bundles of collagen and elastic fibrils, which form an intricate network around alveoli and the alveolar septum. The main interstitial cell is the fibroblast, which is responsible for the synthesis and maintenance of the collagen and elastic fibres (Fig 109a). Other cells found in this space and macrophages and occasional wandering blood cells such as lymphocytes and granulocytes.

NERVOUS TISSUE

The nerve cell or neuron is differentiated for the conduction and transmission of impulses. Despite these properties, the cytoplasm of neurons contains organelles and inclusions typically found in most other cells. The cell body or perikaryon may emit one or two short outgrowths or dendrites, which carry nerve impulses centripetally. Dendrites have irregular surfaces and may penetrate between epithelial cells, where they receive stimuli from the surface or from other receptors. The axon is usually a long process which carries impulses centrifugally to the next neuron, as well as to muscle cells, and gland and surface epithelial cells. Neurons are adapted to their specialized functions by means of the different types of outgrowths. Some neurons have only the axon,

i.e. are monopolar, others have one dendrite and one axon, i.e. are bipolar, and still others have many outgrowths, i.e. are multipolar. In invertebrates most neurons are monopolar. The dendrites of many bipolar neurons bear non-motile, usually sensory cilia, which originate from basal bodies and project above an epithelium (see page 135). These cilia may also have centrioles associated with them, Centrioles do not occur in fully differentiated neurons, except in association with the basal body of a cilium. However, paired centrioles are present in neuroblasts. The absence of centrioles after embryonic life may correlate with the fact that differentiated neurons do not divide but remain in permanent interphase throughout the entire life of an organism.

Vertebrate neurons and their processes are surrounded by special cells and their sheet-like

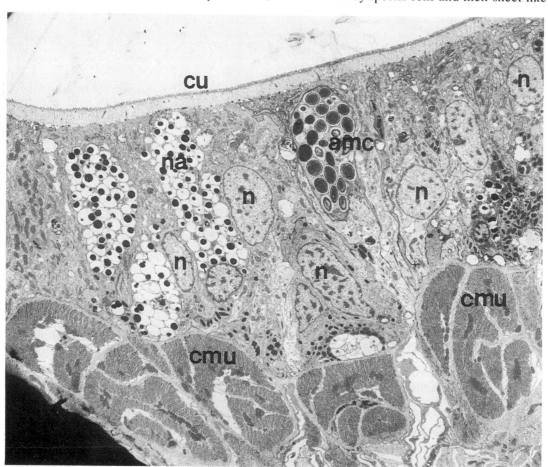

Fig. 99(a) – Longitudinal section through the epidermis of the annelid *Lumbricus reynoldsoni*. The secretory cells are not restricted in their position. × 3 000. (Dr K.S. Richards.)
amc, acid mucous cell; cmu, circular muscle; na, non-acid mucous cell.

extensions. These cells are called Schwann cells, and satellite cells in the peripheral nervous system. The basic function of these cells is to protect and support the neuron and to help in the functions fundamental to the initiation and propagation of nerve impulses. Dendrites and the perikaryon are surrounded by such cells, whereas their peripheral surface is covered by a thin external (basal) lamina. The cytoplasmic encasement of axons by Schwann cells can either be simple or complex. In the former, axons are contained within invaginations of the Schwann cell cytoplasm and are referred to as non-myelinated nerve fibres (Fig 111a). However, in the latter, the innermost portion of the Schwann cell cytoplasm is wrapped around the axon. During the wrapping process, cytoplasmic sheet-like extensions flatten out and disappear to give a 'Swiss roll' configuration made up of concentric layers of lipoprotein membranes, known as a myelin sheath. The axon is then said to be myelinated (Fig 111b).

Peripheral nerves are covered by a loose connective tissue, the epineurium. Nerves are composed of bundles of both non-myelinated and myelinated axons or nerve fibres, the perikarya of which lie in the central nervous system. The majority of cell nuclei present in nerves belong to the Schwann cells which cover the nerve processes (Fig 111a). The Schwann cell is, in turn, surrounded by a small quantity of loose connective tissue, the endoneurium, whereas individual nerve axons are grouped to form nerve fascicles, surrounded by a sheath of dense connective tissue, the perineurium (Fig 111).

Nerve fibres, or axons, contain thin long mitochondria, microtubules (neurotubules) and neurofilaments, as well as vesicular and tubular profiles of agranular ER and a few multivesicular bodies. Ribosomes and granular ER do not occur in the axoplasm. Neurofilaments predominate in large axons, but the ratio of neurofilaments to microtubules decreases in smaller axons. Axons vary in diameter from 1 to 20 μm and it has been established that those with a large diameter conduct nerve impulses faster than smaller diameter axons.

Invertebrate nerves are surrounded by a thin external (basal) lamina, in the outer reaches of which collagen fibrils are normally found embedded. Each nerve usually comprises

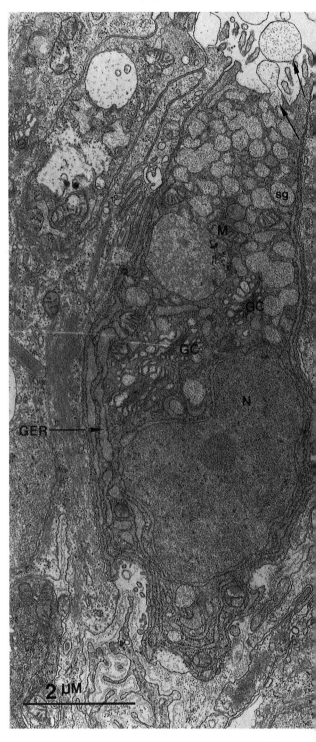

Fig. 99(b) – A mucous cell in the collar of the polychaete, *Spirorbis spirorbis*. Release of secretory mucous granules (sg) is by apocrine secretion (arrows). × 19 000. (Dr A. Bubel.)

159

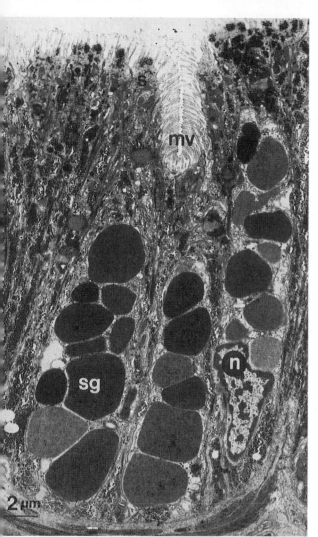

numerous non-myelinated axons interlaced by cytoplasmic processes from glial or supportive cells, most of which are scattered along the periphery of the nerve. Glial cells are generally irregularly outlined, containing, within the cytoplasm of the cell body, characteristically numerous electron-dense or moderately electron-dense membrane-bound granules, or gliosomes. The cytoplasmic processes, in contrast, contain very few organelles, but are mainly filled with glial filaments, approximately 5 nm in diameter. At the periphery of a nerve, glial cells are covered by a thin external (basal) lamina. In peripheral regions of a nerve, immediately beneath the glial cell membrane, hemidesmosome-like structures are evident in which glial filaments appear to terminate (Fig 112a). In the nerves of some invertebrates, glial or sheath cells are apparently absent (Fig 113). Axons are directly contiguous with one another, being separated by a narrow space, approximately 15 nm wide and contiguous with glial processes that penetrate the nerve (Figs 112b, c, d). The axons in

Above
Fig. 99(c) – Mucous cell of the bivalve, *Cardium edule* with fibrous contents situated in the mantle epidermis. × 7 200. (Dr A. Bubel.)

Right
Fig. 99(d) – Mucous cell with granules of differing electron density situated in the mantle of the bivalve, *Nucula sulcata*. × 3 400. (Dr A. Bubel.)

Far right
Fig. 100 – The submaxillary gland of the rat. The cell is packed with secretory granules (sg), many of which are partly coalesced. × 9 500. (Dr T. Jenkins.)

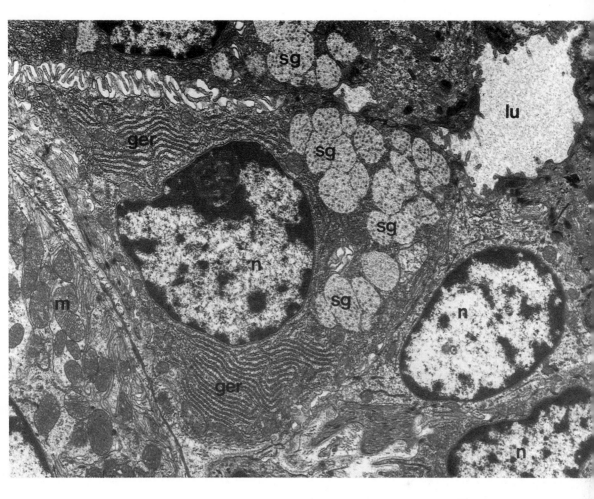

some nerves appear to be segregated by glial cell processes into sizeable groups composed of 50 or more small axons between 0.1 and 0.5 μm in diameter, amongst which a few larger axons, from 1 to 7 μm in diameter, are scattered. For the most part, large axons are confined to the more central regions of a nerve. Mitochondria, ∝-glycogen rosettes, microtubules and neurofilaments are present in the axoplasm of all axons (Fig 112b, c). Some axons contain a mixture of a large number of clear vesicles, 20–30 nm in diameter, and a few larger dense-cored vesicles, 50–125 nm in diameter (Fig 112b, c). Axons containing such vesicles are, as a rule, confined to inner regions of a nerve (Fig 112b, c).

A specialized membraneous contact between the axon ending of one neuron and a dendrite or perikaryon of another neuron is called a synapse. At the terminations of axons, swellings occur which contain a varied number of small mitochondria and clear vesicles, called synaptic vesicles, which are assumed to contain neurotransmitters. In some invertebrate nerves, some synapses appear to be formed *en passant* rather than at specialized endings (Fig 114). Also, although synaptic vesicles are usually clear, some dense-cored examples, are found in some nerves. The specialized regions of membraneous contact consist of the pre-synaptic membrane of the axon, the post-synaptic membrane of the dendrite or perikaryon and the intermembraneous synaptic cleft, averaging 20–30 nm wide. Dense filamentous material is accumulated at the cytoplasmic aspect of the postsynaptic membrane, the subsynaptic web. This gives the synaptic region a polarized appearance. Multiple synapses, usually involving two pre-synaptic axons and a single postsynaptic profile, are common in some invertebrate nerves (Fig 114). This kind of synapse is

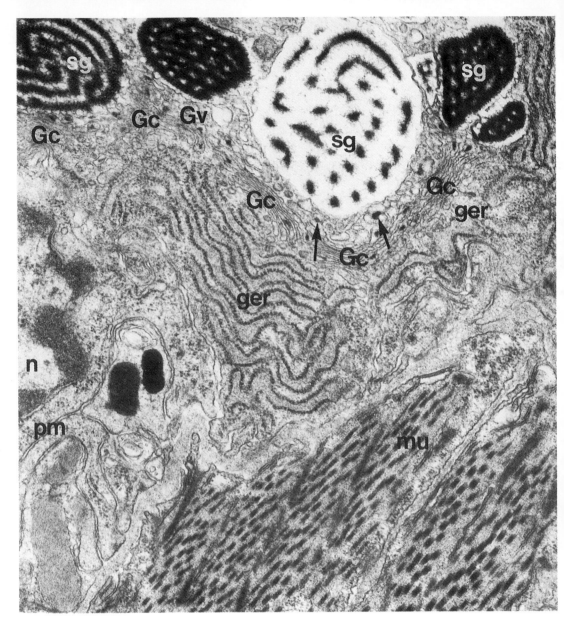

Above
Fig. 101(a) – The base of an acid mucous cell of the annelid, *Lumbricus georgiensis* with a prominent Golgi complex. The secretory mucous granules (sg) originate towards and at the maturing face of the Golgi complex (GC) (arrows). × 34 000. (Dr K.S. Richards.)

Right
Fig. 101(b) – Diagram of Fig. 101(a).

referred to as a chemical synapse, since the chemical transmitter of the synaptic vesicles is assumed to be extruded into the synaptic cleft for action on the postsynaptic membrane, thereby resulting in the transmission of the nerve impulse across the synaptic cleft. In lower vertebrates and some invertebrates, electrical synapses are found. These resemble gap junctions. Electrical resistance between the cells is low at these junctions, and ions diffuse freely through these points. Thus, no chemical transmitters are required for passing

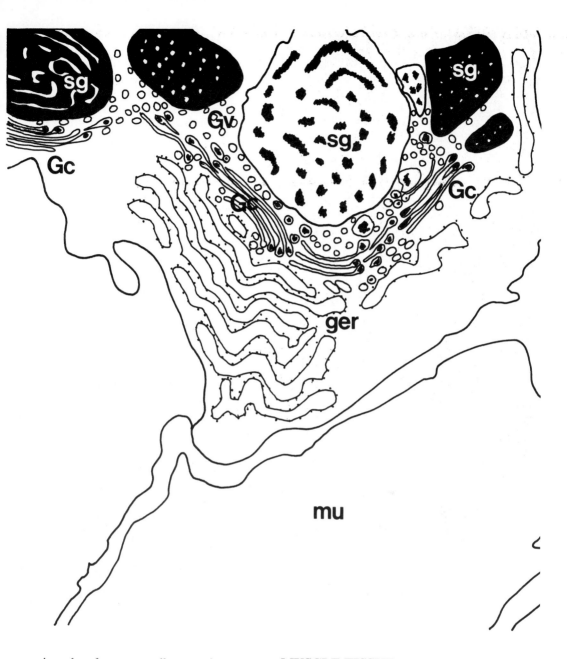

nerve impulses from one cell to another.

In invertebrates, nerve branches composed of closely packed axons and associated glial cells are found to terminate between epidermal cells (Figs 115a, b), at bases of gland cells, between myoepithelial cells of some blood vessels (Fig 116a, b) and between connective-tissue cells (Fig 116c). The axons that are presumed to constitute the nerve endings in such regions contain differing amounts of clear and dense-cored vesicles.

MUSCLE TISSUE

Muscle tissue is made up of elongate or spindle-shaped cells bundled together by connective tissue elements. Each bundle of muscle cells is known as a fascicle. Many fascicles make up a muscle. The contractile properties of muscle cells is determined by the presence of the protein actinomyosin. This is a combination of the contractile filamentous proteins actin and myosin. On a functional basis, three

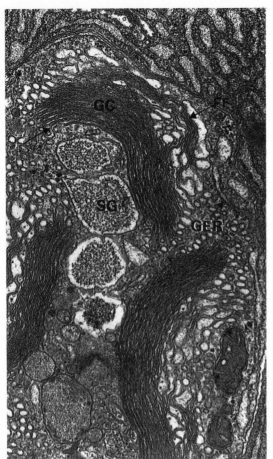

types of muscle tissue can be identified in vertebrates, namely: (1) striated skeletal voluntary, (2) striated cardiac involuntary, and (3) smooth involuntary. Unlike in vertebrate muscle tissues, which are readily identifiable as striated and smooth on the basis of their structure, in invertebrate muscles this primary distinction is not always easy to make.

Striated muscle

The muscle cells are bordered by a cell membrane, the sacrolemma, which encloses the cytoplasm or sacroplasm, that is mostly occupied by longitudinally arranged bundles of myofibrils. The very small amount of extra-fibrillar sacroplasm left contains peripherally situated nuclei, mitochondria, lipid droplets, glycogen particles, and an extensive agranular ER or sacroplasmic reticulum. Ribosomes and granular ER are normally absent. Synchronous insect flight-muscle cells are radially symmetric with mitochondria arranged in a spoke-like

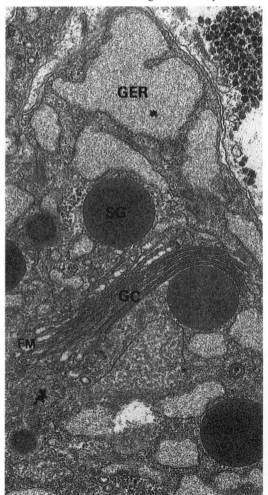

Above

Fig. 101(c) – A thorax gland cell of the polychaete *Pomatoceros lamarkii* with secretory granules with fibrous contents (SG) which appear to originate at the maturing face cisterna (arrow) of the Golgi complex. In the vicinity of the Golgi complex there is a close association between a granular endoplasmic reticulum (GER) and the forming face of a Golgi cisterna (FF). × 16 000. (Dr A. Bubel.)

Right

Fig. 101(d) – A thorax gland cell of *P. lamarkii* containing homogeneously electron-dense secretory granules (SG). There is a close relationship between the Golgi complex (GC), the secretory granules and the dilated, irregularly outlined, granular endoplasmic reticulum profiles (GER). × 16 000.

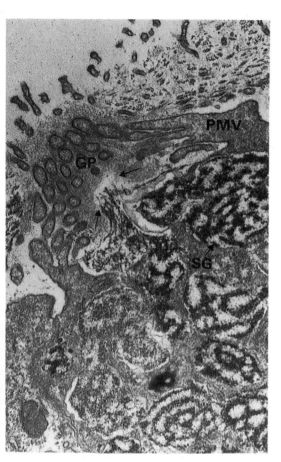

Left
Fig. 102(a) – A thorax gland cell of *Pomatoceros lamarkii* containing secretory granules (SG) with reticulately arranged electron-dense fibrous strands being released (arrows) through a gland cell pore (GP) surrounded by pore microvilli (PMV). × 20 000. (Dr A. Bubel.)

Below
Fig. 102(b) – The apex of an orthochromatic mucous cell (MC) of *Eisenia foetida,* from which a heterogeneous globule of material is released through a pore (GP) around which are pore microvilli. × 7 000. (Dr K.S. Richards.)

fashion around a central tube containing the nuclei (Fig 117). However, asynchronous muscle cells, unlike synchronous cells, contain mitochondria and nuclei distributed throughout the cells in a non-symmetric arrangement.

Myofibrils are made up of regularly arranged myofilaments, which are, in turn, made up of a combination of the contractile proteins, myosin and actin. The myofibrils characteristically possess transverse striations. These result from myofibrils having alternating dark (A-band) regions and light (I-band) regions which are in register with those of adjoining myofibrils. Further, the light segment is bisected by a thin dense area, the Z-line, whereas the centre of the dark segment is occupied by a less dense zone, the H-band. The striations consist of the repetition of a fundamental unit, the sacromere. The region extending between two Z-lines forms the sacromere. The musculature in swimming coelenterate medusae, brachiopod tentacles, crayfish abdominal muscles and insect flight-muscles consists of striated

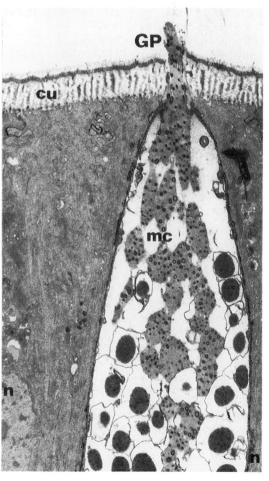

165

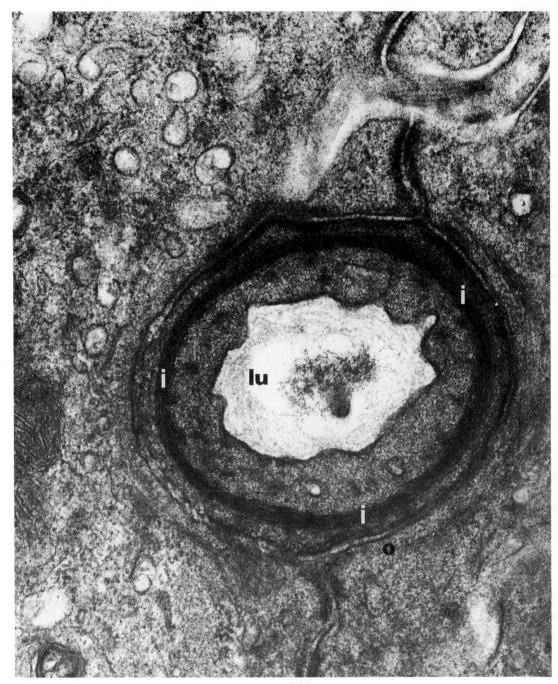

Above

Fig. 102(c) – A slightly oblique transverse section through the pore and sphincter region of an acid mucous cell of *Lumbricus mirabilis,* revealing an inner sphincter (i) within the mucous cell and portions of the outer sphincter (o) within the adjacent supporting cells. × 66 000. (Dr K.S. Richards.)

Right

Fig. 103(a) – A longitudinal section of a chloride cell from the pupfish, *Cyprinodon variegatus* adapted to 100% sea water. The apical surface of the cell invaginates to form an apical crypt containing amorphous mucus material (arrow), and the cytoplasm contains a large population of mitochondria (m). x 11 000. (Dr K.J. Karnaky, Mount Desert Island Biological Laboratory.)

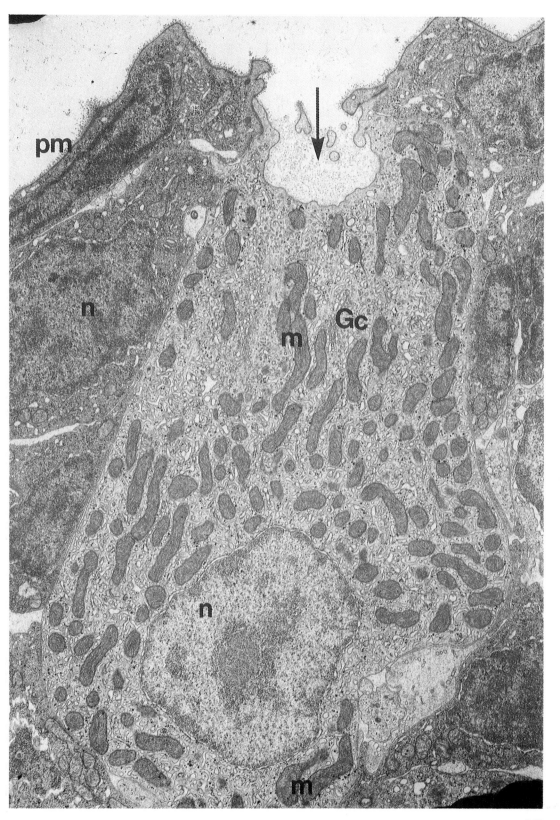

167

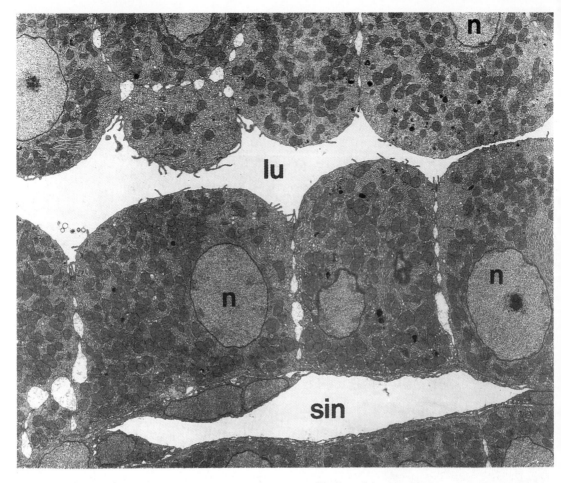

Above
Fig. 103(b) – Several cuboidal cells of the kidney distal segment of the lamprey, *Lampetra japonica*. × 3 400. (Dr T. Hatae, Kagawa Medical School.)

Right above
Fig. 104(a) – A section from a kidney cell from the distal segment of the kidney of *L. japonica*. The cytoplasm contains an extensive reticular network of cytoplasmic tubules (t) and numerous mitochondria (m). In some regions, the cytoplasmic tubules run parallel to form bundles (*). × 6 300. (Dr T. Hatae.)

Right below
Fig. 104(b) – The cytoplasmic tubules from the kidney cells of *L. japonica* contain linear arrays of electron-dense materials on their luminal surfaces. These lie at an approximately 45° angle to the long axis of the tubules. × 47 000. (Dr T. Hatae.)

myofibrils with A-, H-, I- and Z-bands that correspond to those of vertebrate musculature (Figs 118a, b, c, 119a, b). The sacromere length in brachiopod striated myoepithelial cells is 2.5 μm, whereas in crayfish striated fast-receptor abdominal muscle it is 3.3 μm.

Each myofibril is composed of two types of myofilaments, namely, thick and thin. The thick myofilaments are made up of myosin and are 10 nm in diameter and 1.5 μm long, whereas the thin myofilaments are made up of actin and are about 5 nm in diameter and 1 μm long. In brachiopod and crayfish striated myofibrils, the thick filaments are between 20 and 22 nm in diameter and between 2.0 and 2.3 μm long, whilst the thin filaments are between 5 and 7 nm in diameter and 1 μm long (Fig 120a, b). In the middle of a thick myofilament there is a light zone, the M-band. However, no M-band is visible is some invertebrate thick myofilaments.

Thick myofilaments are concentrated in the

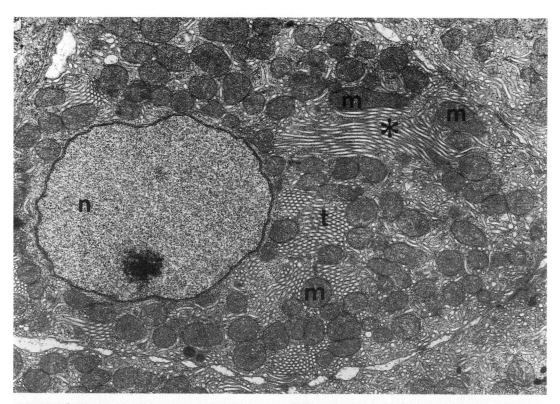

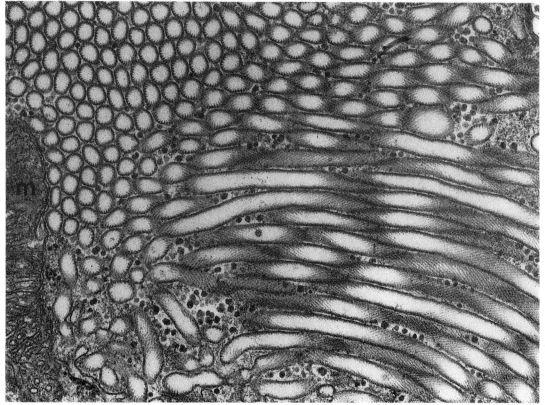

169

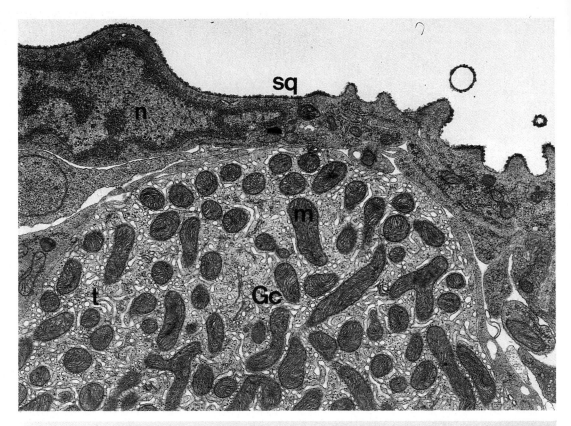

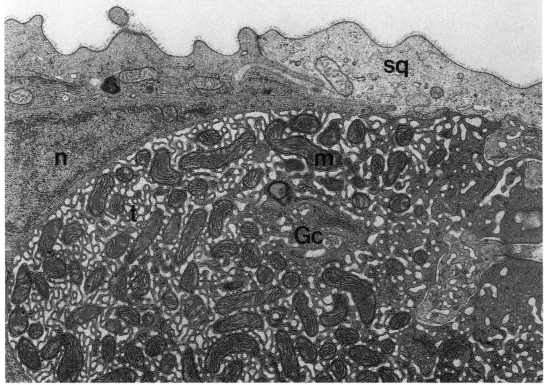

Left above

Fig 105(a) – A chloride cell from *Cyprinodon variegatus* adapted to 100% sea water. The extensive branching tubular system (t) and mitochondria (m) occupy a large portion of the cytoplasm. × 15 000. (Dr K.J. Karnaky.)
Sq. squamous epithelial cells.

Left below

Fig. 105(b) – A chloride cell from *C. variegatus* adapted to 200% sea water (compare with Fig. 105(a)). Two major changes are apparent in the cells. First, the tubular system (t) appears to occupy a much larger proportion of the cytoplasm than in cells adapted to 100% sea water. Second, the mitochondria (m) are small and much more numerous in 200% adapted cells. × 15 000. (Dr K.J. Karnaky.)

Right

Fig. 106(a) – A Malpighian tubule cell from a fifth-stage larva of the insect *Calipodes ethlius*. A homogeneously stained basal lamina (b1) which is attached to basal cytoplasmic processes (cp) separates the cell from the haemolymph (h). × 3 200. (Dr. J.S. Ryerse,
St. Louis Univesity School of Medicine.)
lu, tubule lumen.

Below

Fig. 106(b) – The basal lamina of the Malpighian tubule cell of *C. ethlius* consists of an electron-lucent matrix of granules and fine filaments within which electron-dense fibres are embedded. The matrix is connected to the basal tips of the cell processes (cp) by filaments which emanate from hemidesmosomes (arrows). × 54 000. (Dr J.S. Ryerse.)

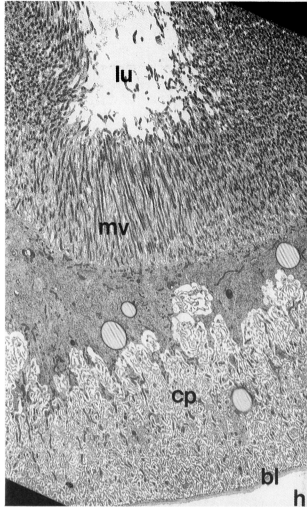

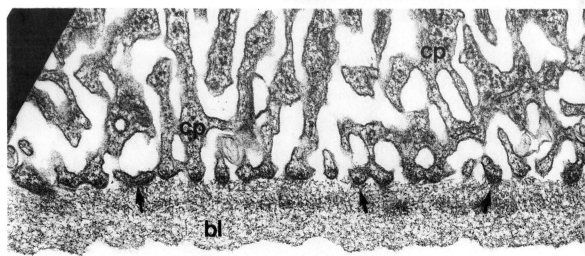

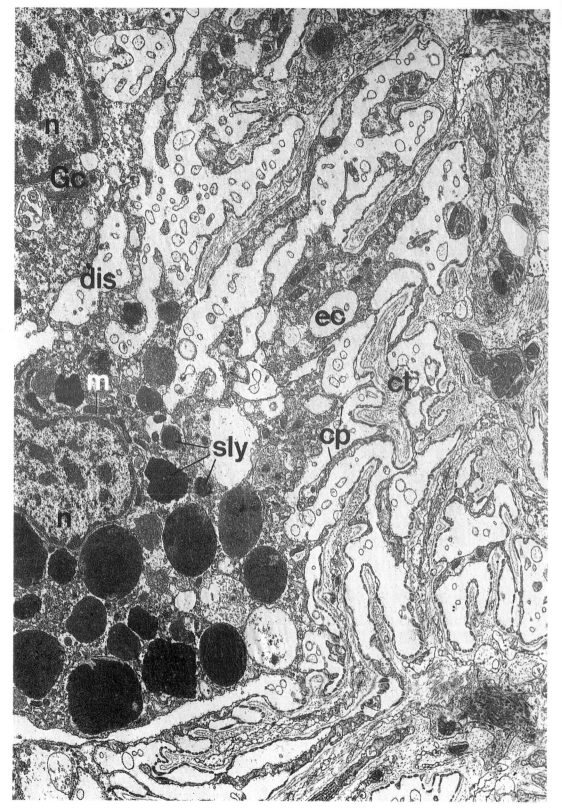

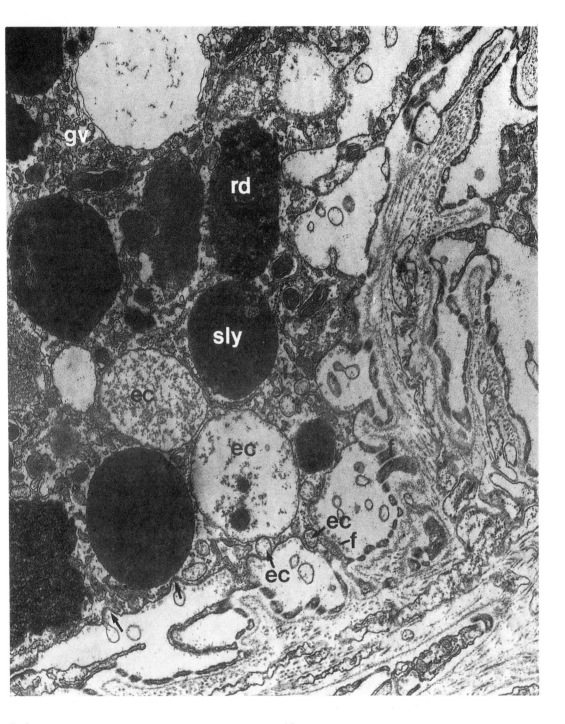

Left
Fig. 107(a) – The basal region of a pericardial cell of the mollusc *Mytilus edulis*. This region of the cell contains cytoplasmic processes (cp) and endocytotic vesicles (ec). × 17 500. (Dr A. Bubel.)

Above
Fig. 107(b) – Lining the basal cytoplasmic membranes and the lumina of the endocytotic vesicles of the pericardial cells of *M. edulis* is adsorbed extracellular filamentous material (f). There is also a close association between Golgi vesicles (gv) and endocytotic vesicles. × 40 000. (Dr A. Bubel.)

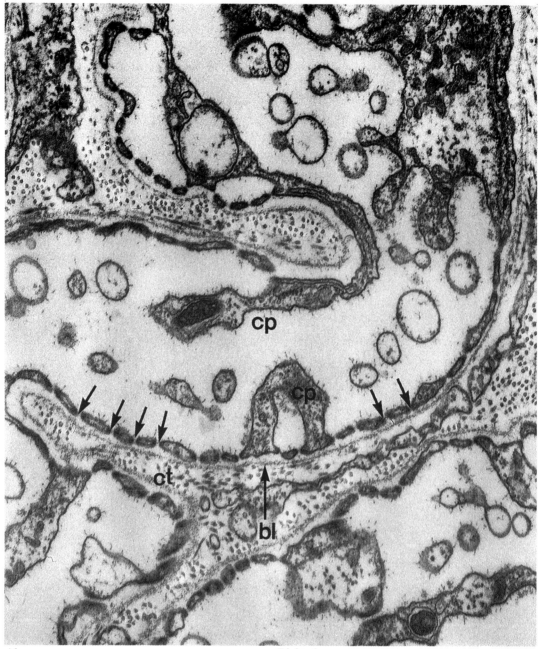

Above

Fig. 107(c) – In the basal region of the pericardial cells of *M. edulis* is a densification of the cytoplasm on apposed lateral cytoplasmic processes. The termination of the cytoplasmic processes gives an attenuated fenestrated appearance to the basal region (arrows). × 48 000. (Dr A. Bubel.)

Right above

Fig. 108(a) – The parenchyma (respiratory zone) of the lung of the bat, *Miniopterus minor* composed of alveoli (a) and blood capillaries (bc). Alveolar macrophagues (ma) are evident passing through interalveolar pores (arrows). × 4 800. (Prof. J.N. Maina, University of Nairobi.)
Edn, endothelial cell nucleus; er, erythroctye.

Right below

Fig. 108(b) – Diagram of Fig. 108(a).

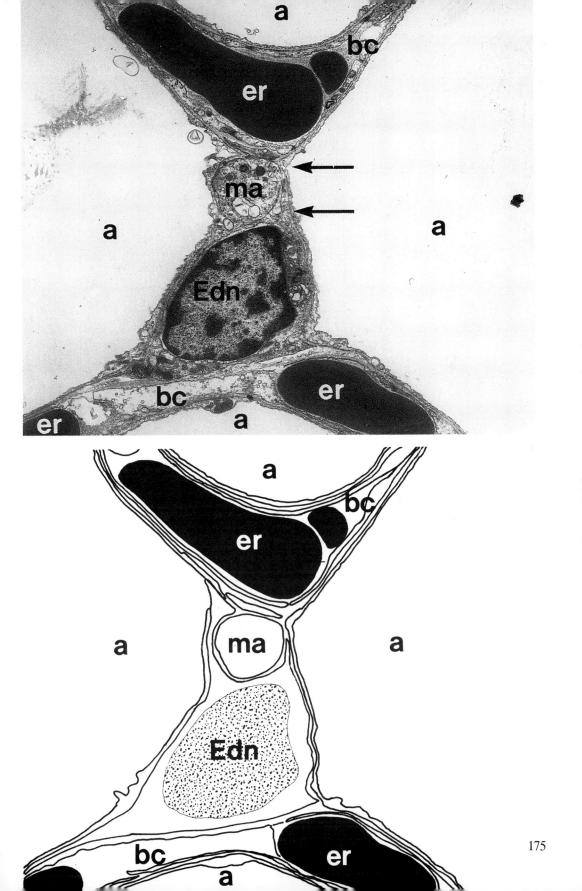

175

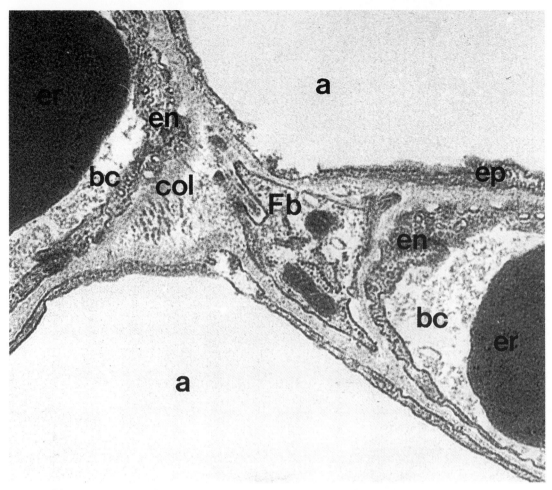

Fig. 109(a) – Two blood capillaries lying adjacent to each other in the lung of *M. minor*. A fibrocyte (Fb) lying close to collagen fibrils (col) is evident. Also, the layers of the blood-gas barrier, epithelium (ep) and endothelium (en) is apparent. × 14 000. (Prof. J.N. Maina.)

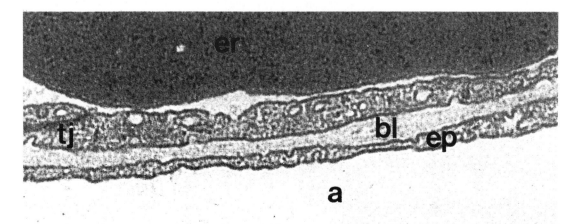

Fig. 109(b) – The components of the blood-gas barrier in the lung of *M. minor*. x 22 500. (Prof. J.N. Maina). tj, endothelial cell tight junction.

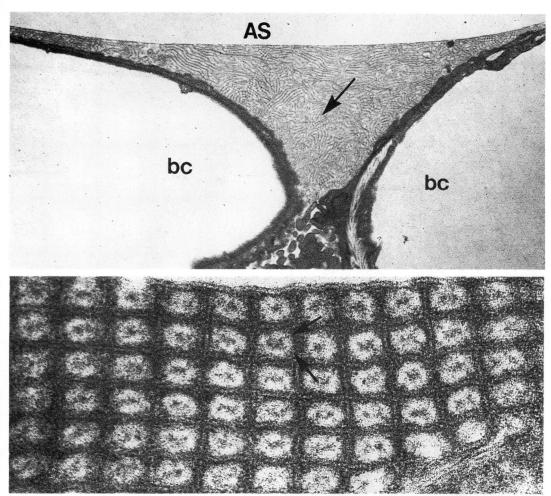

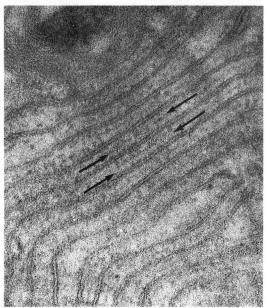

Above top
Fig. 110(a) – Tubular myelin figures (arrow) present in an alveolar cleft in the lung of the rat. × 63 000. (Dr H-J Beckmann, University of Munster.) bc, capillary; AS, air space;

Above centre
Fig. 110(b) – A cross-section through the tubular myelin figures found in rat lung. Tubules with a nearly quadratic profile containing only one central core can be differentiated from rectangular ones with two cores. The cores are connected with the corners of the tubules through filamentous projections (arrows). × 300 000 (Dr H-J. Beckmann.)

Right
Fig. 110(c) – A longitudinal section through small myelin tubules (diameter less than 40 nm). A filamentous central core runs through the matrix of the tubules (arrows). × 154 000. (Dr H-J. Beckmann.)

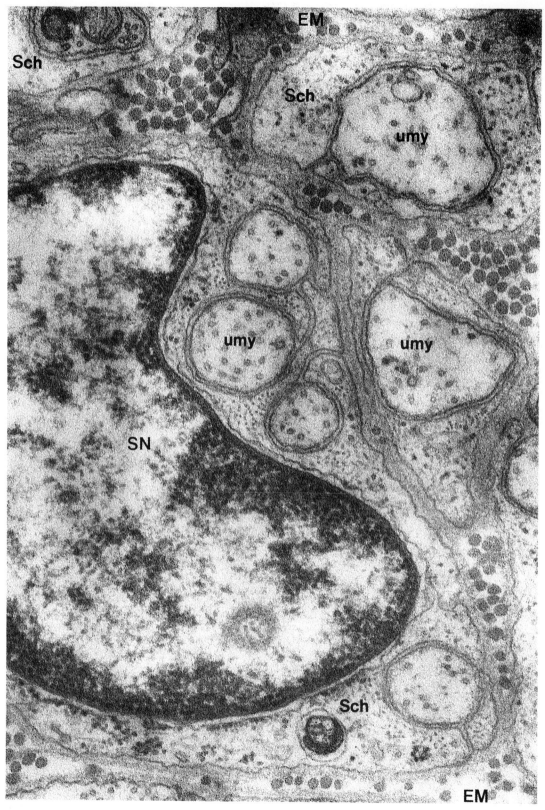

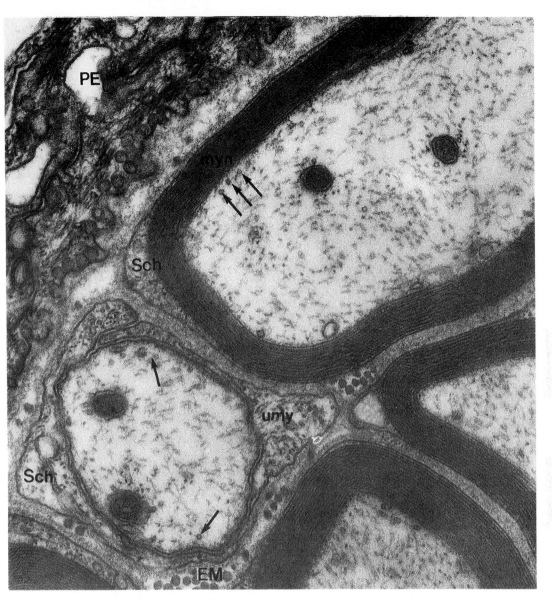

Left
Fig. 111(a) – A portion of a dorsal root central to a spinal ganglion from *Lacerta muralis* revealing unmyelinated axons (umy). × 75 000 (Dr E. Pannese.)
EM, endonerium; SN, Schwann cell nucleus.

Above
Fig. 111(b) – A portion of a dorsal root central to a thoracic spinal ganglion from the lizard *Lacerta muralis*. Microtubules (arrows) are present in the axoplasm of both unmyelinated (umy) and myelinated (myn) axons. × 60 000. (Dr E. Pannese, University of Milan.)
EM, endoneurium; PE, perineurium; Sch, Schwann cells;

A-band, whereas thin myofilaments originate from both sides of the Z-line and extend in opposite directions to form the I-band. The thin filaments pass between the ends of the thick myofilaments within the A-band also. The degree of interdigitation of the thin and thick myofilaments depends on the degree of contraction and relaxation of the sacromere. In the relaxed condition, the I-band contains only thin filaments, the H-band contains only thick filaments, and within the A-band the thick and thin filaments overlap. According to the present understanding of the contraction process, the thin filaments of the I-band slide

179

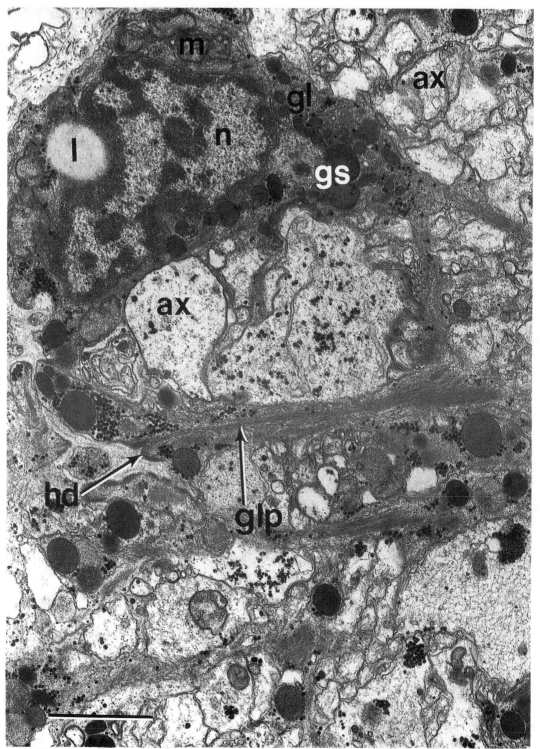

Fig. 112(a) – A cross-section through the peduncle nerve of Pomatoceros lamarkii. The nerve comprises numerous unmyelinated axons (ax) interlaced by cytoplasmic processes (glp) from glial cells (gl). Glial cells contain bundles of filaments and homogeneously electron-dense or fibrous membrane-bound granules (gliosomes) (gs). × 5 800. (Dr A. Bubel.)

between the thick filaments of the A-band.

The interdigitation of thick and thin myofilaments is more clearly seen in cross-sections of myofibrils. In vertebrate muscle cells, the thick myosin filaments are arranged in a hexagonal pattern, about 45 nm apart with six actin thin filaments grouped around each thick myofilament. Each thin filament is shared by three thick filaments. The hexagonal pattern is different in insect flight-muscle. Here, each thin filament is equidistant from two thick filaments and each thick filament appears surrounded by 12 thin ones (Figs 121a,b, 122a,b). In the fast abdominal crayfish muscle, both thick and thin myofilaments are arranged in a highly regular hexagonal pattern in which each thick filament is surrounded by six thin filaments (Fig 120b). However, in brachiopod striated myoepithelial cells, each thick filament is surrounded by 12 thin fila-

ments (Fig 118b). In muscle cells, the higher the organization of the myofilament field the lower the thin : thick myofilament ratio. For example, in crayfish fast muscle cells, the thin : thick ratio is 3 : 1, whereas in brachipod striated myoepithelial cells the ratio is 6 : 1.

Mitochondria are disposed in longitudinal rows between the myofibrils, often opposite I-bands or Z-lines (Figs 118c, 119b). There are greater numbers of mitochondria in steadily active muscle cells. This is related to the constancy with which a muscle contacts. Similarly, glycogen particles are found in large accumulations between myofibrils, especially at the level of the I-bands (Fig 119b). Golgi complexes are, however, located mainly near the poles of a muscle cell.

The sacroplasmic reticulum (SR) is a modified or special type of smooth agranular ER, consisting of tubules and cisternae arranged

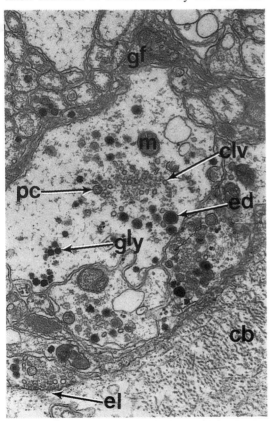

Fig. 112(b) – The peduncle nerve axons of *P. lamarkii* contain a mixed population of vesicles with clear (clv), punctate (pc) and electron-dense (ed) contents. x 25 000. (Dr A. Bubel.)
el, external lamina; glp, glial cell processes containing glial filaments;

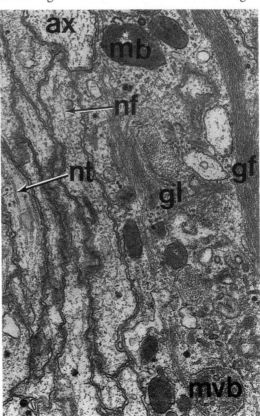

Fig. 112(c) – A longitudinal section through the peduncle nerve of *P. lamarkii* revealing the relationship between a glial cell (gl) and axons (ax) containing neurotubules (nt) and filaments (nf). × 30 000. (Dr A. Bubel.)

longitudinally and transversely around each myofibril. The transverse component of the SR, referred to as the T-system, is apparently continuous at certain points with the sacrolemma. It is the structure best fitted to conduct impulses from the cell surface into the deepest portions of the muscle cell. The abundance and distribution of the SR and its associations with invaginations of the sacrolemma, the transverse tubules (TTs) is extremely important. In mammalian muscle cells, the TTs form terminal cisternae at the level of the A–I junctions, where two terminal cisternae are interposed by a small invagination of the sacrolemma, the TT, to form a triad. Thus, in mammalian muscle cells, each sacromere has two triads located at A–I junctions. In crayfish fast muscle cells, diads and triads are usually located at the level of the A-band. The most prevalent TT–SR coupling observed in arthropods is the diad, consisting of one element of TT paired with one SR cisterna. In synchronous insect flight-muscle, the SR is a vast network of highly fenestrated sheath-like curtains surrounding myofibrils and mitochondria. Usually three to five layers of SR surround both the myofibrils and mitochondria in areas adjacent to the A-band. At the A–I junction, SR–TT couplings are formed by TT, with either one or two sheets of SR to give

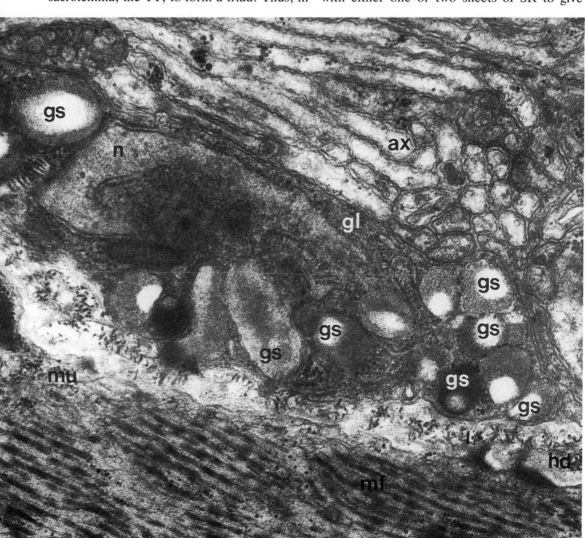

Fig. 112(d) – Glial cells (gl) containing gliosomes (gs) with angular electron-lucent centres from a nerve of the polychaete *Spirorbis spirorbis*. x 37 000. (Dr C.A.L. Fitzsimons, Portsmouth Polytechnic.)

diads or triads, respectively (Figs 122b,c). However, in asynchronous insect flight-muscle, the SR is reduced to a limited tubular network which forms mostly diads with the TT. In both vertebrate and invertebrate muscle cells, diads, triads, tetrads and even pentads have been described. The significance of having diads, triads and tetrads in areas of high SR concentration and reduced numbers of mitochondria, i.e. in axial and terminal regions of the cell, may be related to the function mitochondria have in contraction–coupling. During muscle contraction, mitochondria can serve as a source of calcium ions. Thus, an increase in the amount of SR and SR–TT

junctions would be necessary to ensure adequate release of calcium ions in cell regions where mitochondria are absent. The increased frequency of such junctional complexes could also facilitate the synchronization of contraction throughout the entire volume of a muscle cell. Cardiac muscle cells in mammals come into their closest contact at gap junctions. Mammalian myocardial gap junctions are often extensive, but whether short or long are frequently associated at their cytoplasmic surface with mitochondria, sometimes in their entirety (Fig 9).

In brachiopod striated myoepithelial cells, the SR is not well developed and there is no

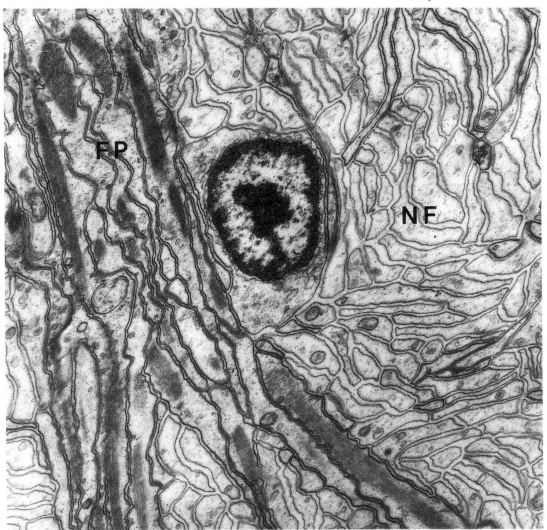

Fig. 113 – A cross-section through the longitudinal nerve of the tube foot of the echinoderm. *Echinus esculentus*. Filamentous processes (FP) from epidermal cells subdivide the nerve into bundles of nerve fibres (NF). × 19 500. (Prof. Dr E. Florey.)

183

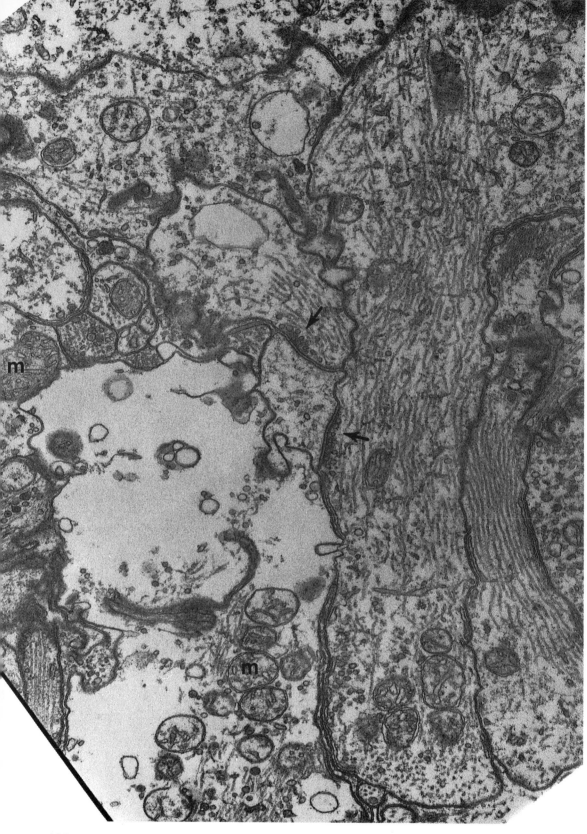

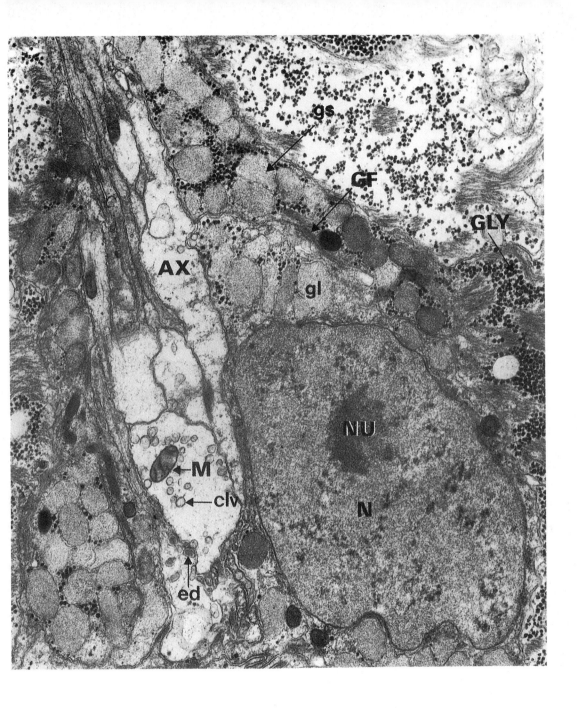

Left

Fig. 114 – A small fibre in the olfactory nerve of the mollusc *Octopus joubini,* receiving two synapses (arrows). × 29 000. (Dr D.G. Emery.)

Above

Fig. 115(a) – A nerve branch composed of axons (AX) and associated glial cells (gl) containing numerous membrane-bound granules, gliosomes (gs), in the basal region of the opercular filament epidermis of *Pomatoceros lamarkii.* The large axons contain clear vesicles (clv) and dense-core (ed) vesicles. × 12 500. (Dr A. Bubel.)

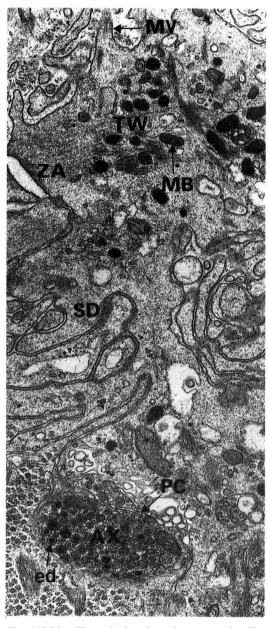

Fig. 115(b) – The apical region of an opercular filament epidermal cell of *P. lamarkii* in which a nerve axon (AX) containing vesicles with electron-dense (ed) and punctate (PC) contents is present in the intercellular space below a septate junction (SD). × 14 500. (Dr A. Bubel.)

evidence of a T-system. Instead the tubular SR frequently forms subsacrolemmal cisternae that are closely applied to the lateral sacro-lemma to form peripheral couplings (Fig 118b).

Microtubules are found in myocardial cells and in vertebrate skeletal muscle cells. They are predominantly orientated along the long axis of the cells between myofibril bundles (Fig 123a). However, they may also be helically arranged around myofibrils and may cross myofibril bundles at, or near, the A–I junction and frequently near the Z–line (Fig 123b,c). An altered arrangement of some myofibrils is found in some regions of muscle cells where microtubules occur (Fig 124). The role of microtubules, although not clear, may be organizational, maintaining structural relation-ships between elements of the sacromere.

In invertebrate muscle cells, e.g. fast cray-fish abdominal muscle and brachiopod striated myoepithelia, patches of electron-dense material are often found on the cytoplasmic aspect of the sacrolemma, which resemble hemidesmosomes. Thin myofilaments appear to insert into such structures.

Smooth-muscle

Smooth-muscle cells in both vertebrates and invertebrates have a more varied organization. They are usually elongate, spindle-shaped cells without transverse striations, arranged with their tapering ends overlapping to form bundles and sheets (Figs 125, 126a–d). The bulk of the cytoplasm is occupied by myofila-ments. In vertebrate smooth-muscle cells there are three types of myofilaments, namely: (1) thick filaments, on average 15 nm in diameter and about 0.6 μm long, containing myosin; (2) a few intermediate-size filaments, on average 10 nm in diameter, length and composition unknown; and (3) thin filaments, on average 5 nm in diameter, containing actin. Invertebrate smooth-muscle cells contain thick and thin myofilaments only. The thick filaments are 40–100 nm in diameter in annelids, 110 nm in diameter in brachiopods; and 25–36 nm in diameter in crayfish, whilst thin filaments are 5 nm in diameter in annelids, 7 nm in diameter in brachiopods, 10 nm in diameter in crayfish; and 9–11 nm in diameter in molluscs. The thin filaments are fusiform and possess a poorly defined axially periodicity, 12–14 nm in anne-lids, 32 nm in brachiopods, and 17 nm in molluscs (Figs 126, 127a,b, 128, 129a,b). The profiles and axial periodicities of the thick filaments are characteristic of paramyosin myofilaments. It seems that the thick filaments may be primarily composed of paramyosin, but that myosin may also be present. The thin filaments are believed to contain actin.

Generally, the arrangement of thick and thin myofilaments in muscle cells appears to be irregular. Thick filaments occur in randomly arranged clusters and often give the impression of being arranged in parallel curved rows (Fig 127a). There appears to be some degree of semi-hexagonal packing of thick filaments over a short range (Fig 128). However, the centre-to-centre spacing is not as regular as it is in striated muscle. In both vertebrate and inverte-brate muscle cells the thick filaments appear to run mainly parallel to the long axes of the cells (Figs 126–130). The thin filaments appear to run parallel to, or to converge on, thick filaments (Fig 130). There is a tendency in some cells for thin filaments to be arranged in orbits around thick filaments, e.g. 12 to 16 in an orbit in molluscan muscle cells (Fig 128) and 12 encircle a thick filament in crayfish muscle (Fig 129b). The somewhat irregular myofila-ment organization and a high thin : thick ratio appears to be characteristic of smooth-muscle cells, e.g. thin : thick ratio in molluscan muscle cell is 30 : 1 and in mammalian smooth-muscle from 12 : 1 to 27 : 1.

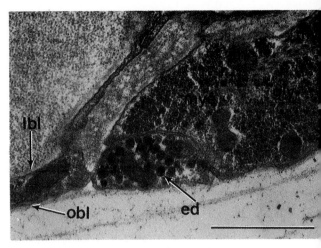

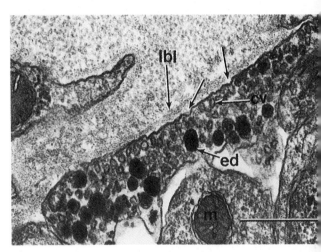

Top
Fig. 116(a) – A nerve axon (AX) packed with electron-dense core (ed) vesicles, situated between the myo-epithelial cells (mye) of the opercular blood vessel of *Pomatoceros lamarkii*. × 29 000. (Dr A. Bubel.)

Centre
Fig. 116(b) – A nerve axon containing clear (cv) and dense core (ed) vesicles contiguous with the luminal basal lamina (lbl) of the opercular blood vessel of *P. lamarkii*. The concavities of the axolemma (arrows) suggest the release of the vesicle contents. × 51 000. (Dr A. Bubel.)

Right
Fig. 116(c) – A nerve axon containing vesicles with dense-core (ed) and punctate (pc) contents enveloped by a cytoplasmic process from a connec-tive tissue cell in the operculum of *P. lamarkii*. × 21 000. (Dr A. Bubel.)

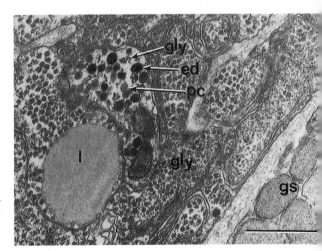

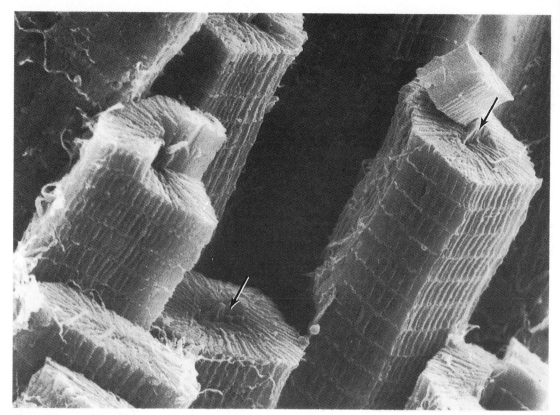

Fig. 117(a) – Insect flight muscle fibres, some with fractured ends, possess a hexagonal symmetry. Nuclei form a central core in the tubular fibres (arrows). × 2 700. (Dr J.B. Delcarpio, Louisiana State University Medical Centre.)

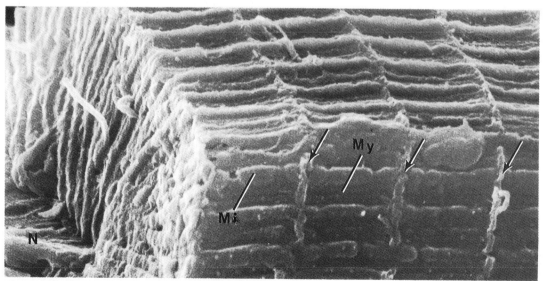

Fig. 117(b) – The typical pattern of mitochondria (Mi) alternating with myofibrils (My) seen in longitudinal and transverse sectional views in a single fractured flight muscle of the insect, *Libellu needhami*. The sacromere boundaries are indicated by transverse raised areas which are believed to consist of sacrolemmal fragments attached to hemidesmosome-like junctions (arrows). × 36 000. (Dr J.B. Delcarpio.)

The sacrolemma has an elaborate and distinctive form: it is very irregular in outline with both large and small invaginations. The form of many of the invaginations suggest pinocytotic activity, there being present several smooth-surfaced vesicles beneath the sacrolemma. The SR is considerably reduced and consists in most cells of a few longitudinal or oblique tubules that run peripheral to the myofilament field. Peripheral tubules and sac-like dilations of the SR form couplings with the sacrolemma and close contacts with surface vesicles. These give rise to dome-shaped protruberances at the muscle-cell surface (Fig 131a). The space between the two membranes is fairly constant at between 6 and 10 nm. No specialized intercellular junctions are obvious between adjacent cells in some muscle tissue. However, areas of membrane contacts may be differentiated into gap junctions (nexuses). Smooth-muscle cells send out a varied number of blunt or short, narrow finger-like processes which make contact in peg-and-socket arrangements and connect to adjacent cells (Fig 131b,c).

Mitochondria are relatively sparsely distributed in most cells. They occur, usually, in central areas of the cell among myofilaments and peripherally where myofilaments are fewer (Figs 126, 128). The Golgi zone is relatively large, but a few lipid droplets, some profiles of granular ER and many free ribosomes are found in the vicinity of the nucleus. Glycogen in ß-particle and ∝-rosette form occur in varying amounts in short rows orientated parallel to the long axes of the myofilaments. Coursing through and parallel to the myofilaments are several microtubules also.

Attached to the cytoplasmic aspect of the sacrolemma at varied intervals, in most cells, are highly dense structures that resemble hemidesmosomes. Internally, randomly distributed dense bodies are found amongst the myofilaments (Figs 126c,d, 128, 130). Thin myofilaments appear to attach to these dense bodies. The dense bodies in vertebrate vascular smooth-muscle are not simply patches of electron-dense material adhering to the inner aspect of the sacrolemma, but are more extensive, wedge-shaped and project into the sacroplasm for up to 1 μm (Fig 130). The distrubution of the dense bodies is considered to be essential as they are anchors for the preservation of organized orientated contraction and for intergrating the contractile activity of smooth-muscle cells.

MUSCLE CONNECTIVE TISSUE

Muscle cells are invested by an endomysium composed of a thin external (basal) lamina and a thin connective tissue layer made up of collagen fibrils (reticular fibrils, fibroblasts, nerves and blood capillaries) (Figs 126, 128). The opposed external laminae of some adjacent invertebrate muscles are close enough

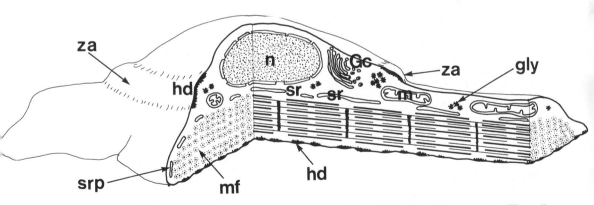

Fig. 118(a) – A diagram of a striated myoepithelial cell of the brachiopod *Terebratalia transversa*. The cell is fusiform and elongated basally along the axis of the myofilament field (mf), but the contractile process is truncated at either end of the perikaryon. The basal part of the cell is occupied by an extensive striated myofilament field embraced peripherally by tubules of sacroplasmic reticulum (sr). The tubular sacroplasmic reticulum forms peripheral couplings (srp) with the lateral sacrolemma. A juxtaluminal zonula adhaerens (za) forms a band around the apex of the cell. Hemidesmosomes (hd) attach the cell basally and transmit the force of contraction to the connective tissue. (After Reed & Cloney, 1977.)

to form cross-bridges. There are also regions that are devoid of basal lamina and collagen fribrils. The membranes of adjacent muscle cells, at such regions, are separated by an intermuscular gap of between 10 and 16 nm wide. In addition, some regions of even closer membrane apposition are found in which adjacent muscle cells are separated by a gap of between 2 and 4 nm wide. Morphologically, such regions are equivalent to gap junctions.

Groups of muscle cells are enclosed by a perimysium, which is thicker than an endomysium, and is made up of bundles of collagen fibrils (some elastic fibrils, fibroblasts, nerves, arterioles and venules). Connective tissue sheets, about 4 μm thick, believed to be a perimysium, are found surrounding bundles of muscle cells in some invertebrates (Fig 126). Finally, a dense regular connective-tissue layer, made up of coarse bundles of collagen fibrils, connects vertebrate skeletal muscle to tendon at the myotendonal junction. In invertebrates a connective tissue layer, forming the myotendon zone or epimysium, connects muscles to specialized 'tendon cells' (Fig 84).

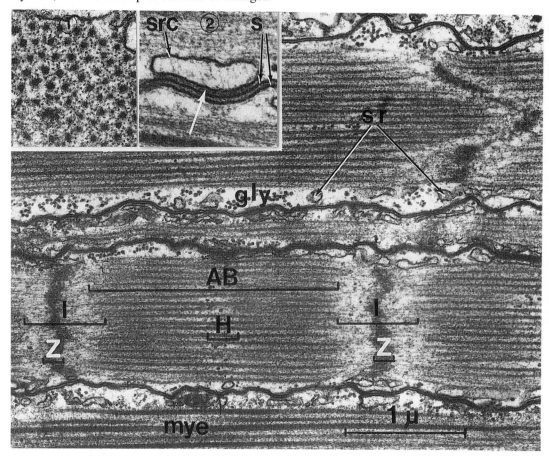

Fig. 118(b) – Longitudinal section through several striated myoepithelial cells of the tentacles of the brachiopod, *T. transversa*. Glycogen particles (gly) are generally restricted to the cell periphery along with the sacroplasmic reticulum (sr), but are found in the myofilament field in the H-band regions where the thin filaments are absent. At the bottom of the micrograph, part of an adjacent smooth myoepithelial myofilament field is visible (mye). × 33 100.
Inset 1. Transverse section of a myofilament field from a striated myoepithelial cell. An orbit of 12 thin filaments surround each thick filament. × 68 000.
Inset 2. Longitudinal section of peripheral coupling in a striated myoepithelial cell. The outer membrane of the subsacrolemmal cisterna (src) closely parallels the sacrolemma (s), and a thin line of electron-dense material is evident within the 15 nm intermembrane gap (white arrow). The lumen of the subsacrolemmal cisterna is electron lucent. × 33 100. (Dr C.G. Reed, University of Washington.)

NEUROMUSCULAR JUNCTIONS

Vertebrate skeletal muscles are innervated by nerve endings or axons terminating in trench-like recesses of the muscle-cell surface, whereas the joining of nerve endings to smooth-muscle cells varies somewhat in that nerve endings either make or do not make membraneous contact. Neuromuscular junctions in invertebrates lack pronounced junctional specialization. The axolemma and sacrolemma are closely apposed with no collagen or basal lamina material interposed between them (Fig 132 a, b). A glial cell is often, but not always, present along the non-junctional surface (Fig 132a, b). The surface of the muscle cell may be partially invaginated to accommo-date an axon, or the axon is enveloped by finger-like cytoplasmic extensions from the muscle cell (Fig 133). The separation between the axolemma and sacrolemma ranges from 5 to 20 nm in width. In general, the junctional portion of an axon contains glycogen particles, a mixture of clear, punctuate and dense-cored vesicles, and an occasional mitochondrion (Figs, 133, 134a–c). The vesicles are believed to contain neurotransmitters. The clear vesicles range from 20 to 30 nm in diamter, whereas vesicles with punctate or dense-core contents enclosed by a peripheral halo are larger and range from 60 to 120 nm in diameter (Fig 133, 134). There is a dual innervation of the anterior byssus retractor muscle of bivalve molluscs which is capable of sustained contrac-

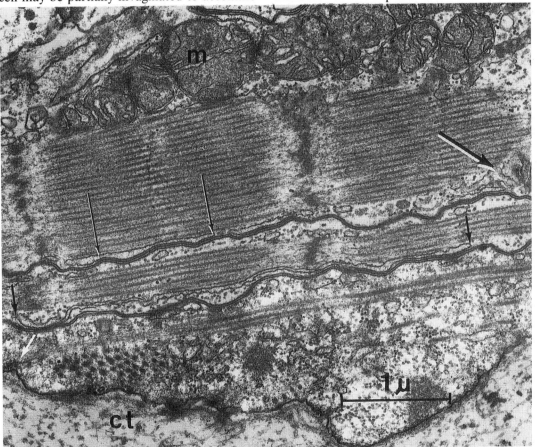

Fig.118(c) – Longitudinal section through a striated myoepithelial cell of *T. transversa*. Tubules of sacroplasmic reticulum arise from meandering longitudinal tubules in the peripheral cytoplasm (large arrow). Occasionally the sacroplasmic reticulum component of a peripheral coupling consists of a long tubule about 4 μm long (between short arrows). The sacromeres are aligned in adjacent fibres. A smooth myoepithelial cell (bottom) contains a small bundle of myofilaments perpendicular to the main myofilament field. The thin filaments in the smooth fibre appear to insert on the basal sacrolemma in densities that resemble hemidesmosomes (white arrow). × 29 500. (Dr C.G. Reed.)

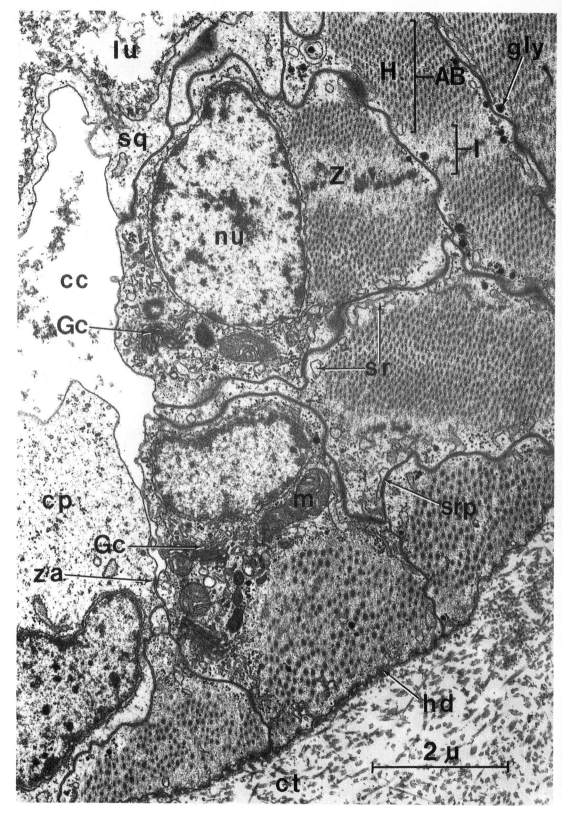

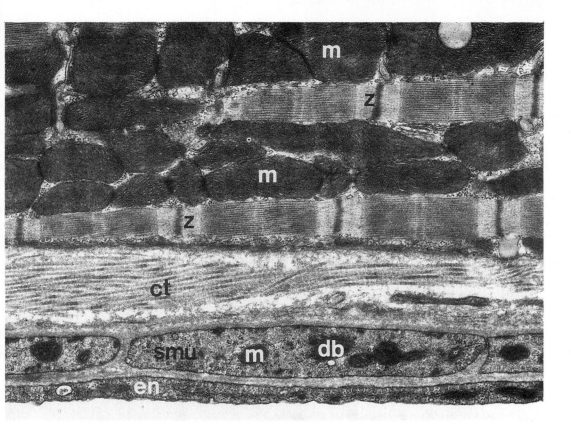

Left

Fig. 119(a) – A slightly oblique section through smooth (bottom) and striated (top) myoepithelial cells of the brachiopod, *T. tranversa*. The myofilament field in each smooth muscle cell consists of much thicker filaments surrounded by many thin filaments. The sacroplasmic reticulum (sr) is more abundant and the peripheral couplings (srp) more frequent in the striated muscle cells. Hemidesmosomes (hd) attach the myoepithelial cells firmly to the connective tissue (lower right). × 18 400. (Dr C.G. Reed.)

cc, coelomic canal; lu, lumen of blood vessel; sq, squamous myoepthelial cell of blood channel;

Above

Fig. 119(b) – Rat myocardium. An arteriole (en) lies parallel to a cardiac muscle cell. In smooth muscle cells (smu), mitochondria (m), and dense bodies (db) are evident. × 24 500. (Dr G. Gabella, University College, London.)

Z, Z lines of myofibrils.

tion. Two nerve endings, which resemble cholinergic and serotonergic types, are distinguished morphologically and by means of fluorescence histochemistry, produce a sustained contraction, and are responsible for muscle relaxation. Axons that contain a large number of clear vesicles closely resemble cholinergic nerve endings in vertebrate skeletal, smooth and cardiac muscle (Fig 134a). In contrast, axons that contain large numbers of dense-cored vesicles and only a few clear vesicles resemble serotonergic axons in other invertebrates (Fig134b, c).

MYOBLASTS

Muscles in vertebrates arise in the embryo as the result of the fusion of several primordial cells: the so-called myoblasts. However, there appears to be no general consensus of opinion with regard to the origin of myoblast cells in invertebrates. In some invertebrates, they are believed to originate either from connective tissue cells or from muscle dedifferentiation (Fig 135). For example, in molluscs it has been

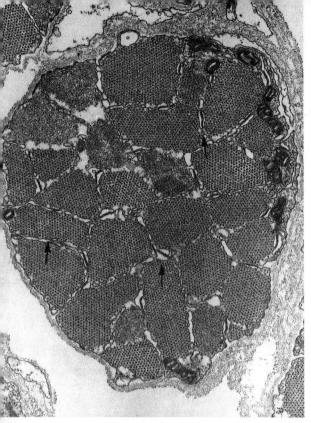

suggested that myoblasts are differentiated from amoebocytes. Myoblasts are undifferentiated cells, with cytoplasm containing many free ribosomes. Myoblasts are located beneath the basal lamina of flight-muscle precursors in some insects. They infiltrate into larval muscle via the channels of the T-system and increase in number by mitosis (Fig 136a, b). Later they help to form the flight muscle. Enigmatic spindle-body cells are found to be the most common cell type in the pedicle coelom of the brachiopod, *Lingula anatina*. These cells contain two types of filaments, namely: thick filaments between 35 and 50 nm in diameter, and thin filaments about 5 nm in diameter. The arrangement and dimensions of the filaments appear similar to those of thick myosin and thin actin filaments in smooth-muscle cells (Fig 137a, b). It is suggested that the spindle-body cells originate from muscle cells.

VASCULAR SYSTEM

Blood vessels

Blood vessels in vertebrates, namely, arteries, arterioles, venules and veins, are composed of three layers: (1) tunica intima, (2) tunica media, and (3) tunica adventitia.

The tunica intima is composed of an innermost lining of endothelial cells, a layer of connective tissue and an internal elastic membrane. The endothelial cells are polygonal, rounded or flat, with the nucleus forming a local bulge into the lumen. From the luminal surface, several processes project; some resemble microvilli. The basal cell region has many small branching foot-like processes descending into the connective tissue (Fig 138a, b). The cells are held together, usually by gap junctions and an occasional macula adhaerens. The edges of adjacent cells overlap in venules, and gap junctions may be found, but often junctional specializations are lacking. In arteries, lymphoctyes may migrate from the blood across the endothelium, enter intercellular junctions and penetrate the basal lamina (Fig 139). There are numerous micropinocytotic vesicles in the cells, providing evidence for metabolic exchange between endothelial cells and their environment. Pinocytotic vesicles may also represent a type of transport vehicle across the endothelial cell. The cells contain small Golgi complexes,

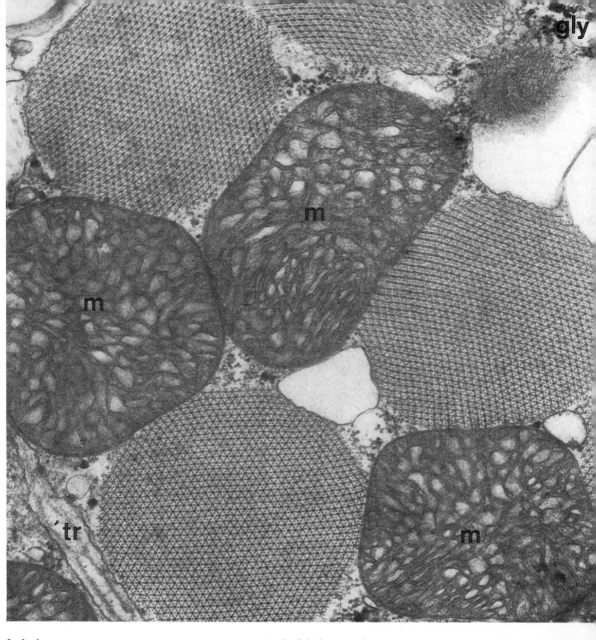

gly

m

m

tr

m

Left above
Fig. 120(a) – Transverse section of a large myofilament bundles from a fast receptor muscle
from the crayfish *Procambarius clarkii*. Polygonal
myofilaments are almost entirely surrounded by a
well-developed sacroplasmic reticulum, which often
makes diadic and triadic contacts with invaginations
of the cell membrane or T-tubules (arrows).
× 12 000. (Dr T. Komuro, Ehime University.)

Left below
Fig. 120(b) – Myofilaments of the fast receptor muscle of *P. clarkii* in transverse section at the level of the
A-band. Both thick and thin myofilaments show
highly organized hexagonal arrangement. The thick
myofilaments appear tubular in transverse section.
× 100 000 (Dr T. Komuro.)

Above
Fig. 121(a) – Transverse section through the flight
muscle of the blowfly, *Calliphora vomitoria*.
Mitochondria (m), glycogen particles (gly) and
trachioles (tr) are intimately associated with the muscle fibrils. × 78 000 (Dr G. Charles.)

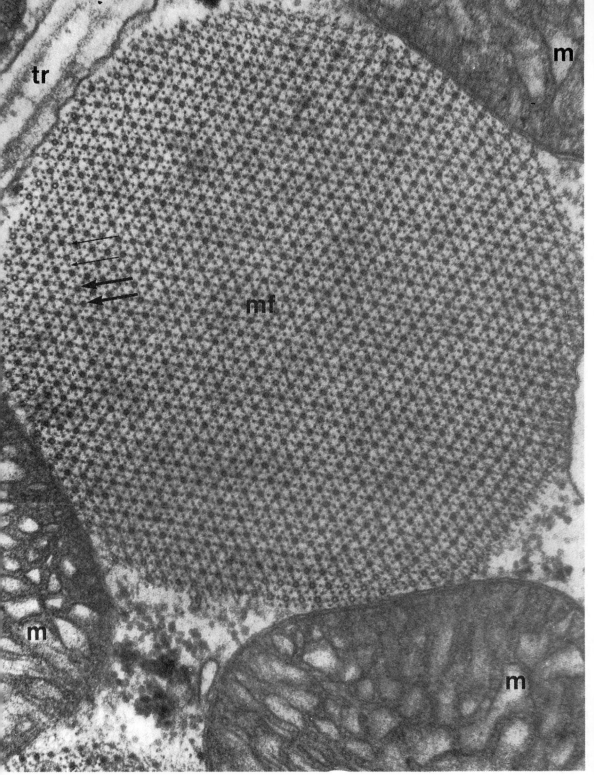

Fig. 121(b) – Transverse section through a flight muscle fibril (mf) of the blowfly, *C. vomitoria*. There is a regular hexagonal pattern formed by thick myosin (large arrows) and thin, actin (thin arrows) filmaments. × 94 000 (Dr G. Charles.)

196

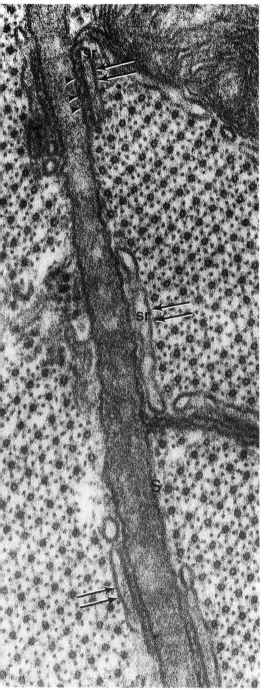

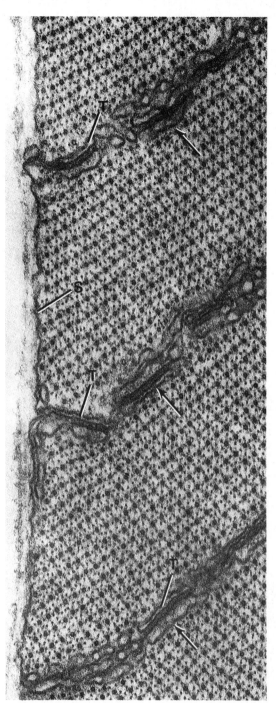

Fig. 122(a) – Transverse section through two adjacent flight-muscle fibres of the insect, *Libellua needhami*. Numerous sacrolemmal junctional complexes are found in these areas (double arrows). There are bridge-like connections between the sacroplasmic reticulum (sr) and sacrolemma (s) (arrow heads) × 96 000. (Dr J.B. Delcarpio, Louisiana State University.)

Fig. 122(b) – Transverse section of insect flight-muscle. Invaginating T-tubules (T) from the sacrolemma (S) meander to form numerous interfibrillar dyads (arrows). × 80 000. (Dr J.B. Delcarpio.)

limited mitochondria, a varied number of specific granules and cytoplasmic filaments. A thin basal lamina or elastic membrane separates the cells from the connective tissue (Fig 138a, b).

The tunica media is composed of concentric elastic membrane sheets and layers of smooth-muscle cells (Fig 138a, b). However, the tunica adventitia is composed of fibrous connective tissue, with an elastic membrane establishing the border between the media and the adventitia (Fig 138). This is the thickest part of the wall in veins and contains some elastic and smooth-muscle cells. Unlike other components of the vascular system, capillaries are com-

posed of a single layer of endothelial cells, a basal lamina and an occasional pericyte (Fig 140a, b). The endothelium in capillaries ranges from 0.08 to 0.2 μm in thickness and may contain regularly spaced fenestrations with diameters of about 20 nm or cytoplasmic holes between 0.5 and 2 μm in diameter.

Unlike the blood vessels of vertebrates, the structure of invertebrate blood vessels is controversial, the existence of an endothelium being frequently disputed. For example, the blood vessel walls of annelids, crustacea, brachiopods and protochordates are composed of an extracellular luminal (vascular) basal lamina that is covered externally by an epi-

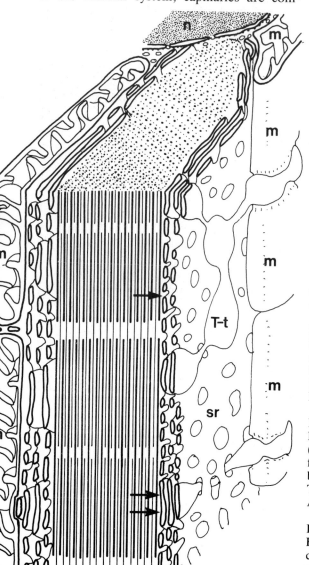

Left
Fig. 122(c) – A diagrammatic reconstruction of the flight muscle of insects (Libellulidae). Invaginating T-tubules (T-t) form a number of couplings with sacroplasmic reticulum (sr) cisternae. T-tubules, which enter at interfibrillar regions or leave mitochondrial troughs near the centre of the fibre, meander through several layers of sacroplasmic reticulum producing interfibrillar diads (single arrow) and triads (double arrow). T-tubules confined to mitochondrial troughs form dyads (single arrow). Rare tetradic couplings (double arrow heads) result when two T-tubules and their associated sacroplasmic reticulum coalesce as adjacent myofilaments abut near the ends or in axial regions of fibres. (After Delcarpio et al, 1983.)

Right above
Fig. 123(a) – Rat soleus muscle. Portions of eight microtubules (Mt) run parallel to the myofilaments in the space occupied by the sacroplasmic reticulum (SR). × 48 000. (Dr J. Cartwright, The Methodist Hospital, Houston.)

Right centre
Fig. 123(b) – Rat soleus muscle. Two microtubules (Mt) curve away from an axial orientation and cross a filament bundle at the junction of the A-band and I-band. × 51 000. (Dr J. Cartwright.)
T, triad of tubules and sacroplasmic reticulum at the A/I junction.

Right below
Fig. 123(c) – Rat soleus muscle. A microtubule (Mt) curves away from an axial orientation and cross a myofilament bundle at the Z-band. × 47 000.

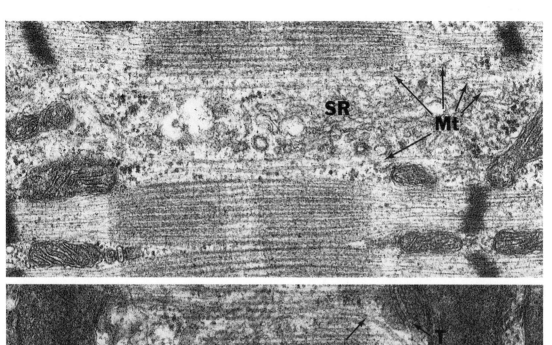

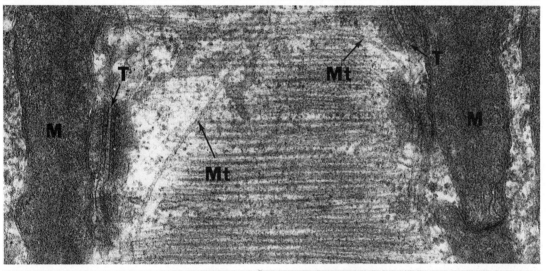

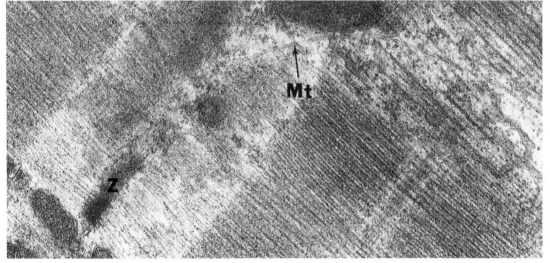

199

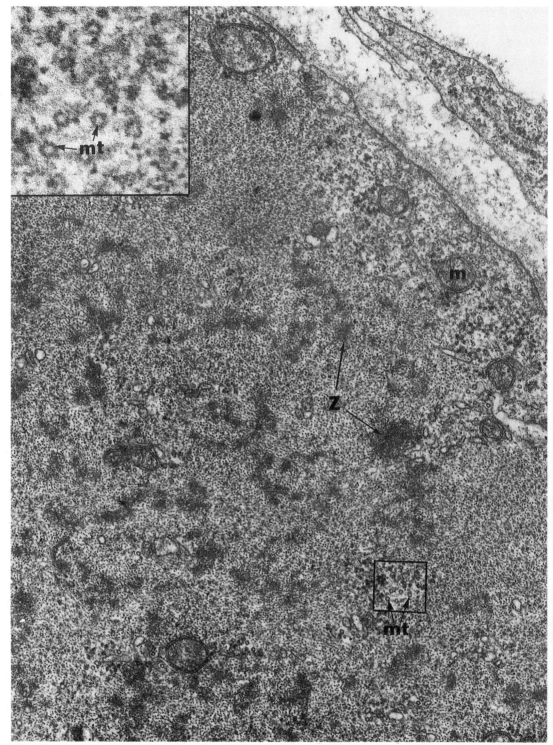

Fig. 124 – A cross-section of an area of rat soleus muscle in which thick and thin filaments are intermingled with patches of Z substance (Z) rather than being organized into distinct bands. × 46 000.
Inset: Area corresponding to the square at higher magnification, revealing details of microtubule (mt) structure. × 167 000. (Dr J. Cartwright.)

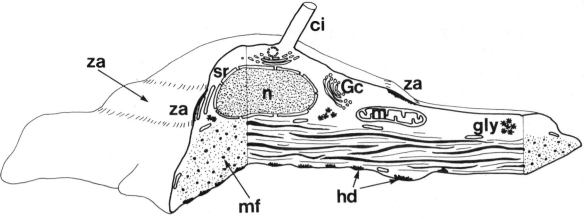

Fig. 125 – A diagrammatic representation of a smooth myoepithelial cell from the tentacles of the brachiopod, *Terebratalia transversa*. A sparse tubular sacroplasmic reticulum (sr) courses over the periphery of the myofilament field (mf), which contains very thick fusiform filaments surrounded by many thin filaments. Occasionally a small bundle of filaments is orientated perpendicular to the filaments of the main myofilament field. A zonula adhaerens (za) rims the apex of the cell and numerous hemidesmosomes (hd) attach the cell basally to the connective tissue. (After Reed and Cloney, 1977.)   ci, ciliary stump;

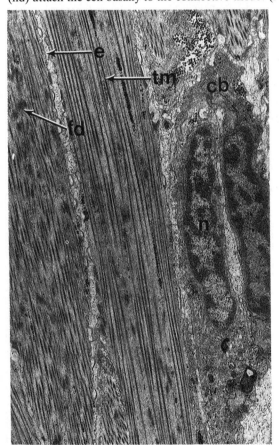

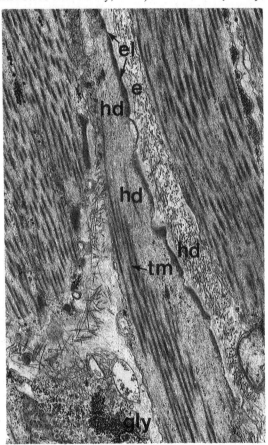

Fig. 126(a) – Longitudinal section of a peduncle muscle of the polychaete *Pomatoceros lamarkii*. × 5 500 (Dr A. Bubel.)
cb, cell body; e, thin collagenous connective tissue sheets (endomysium); fd, free electron-dense bodies; tm, thick myofilaments.

Fig. 126(b) – Longitudinal section of *P. lamarkii* peduncle muscle. A thin collagenous connective tissue sheet (e) (endomysium) separates the muscle fibrils. × 7 000. (Dr A. Bubel.)

201

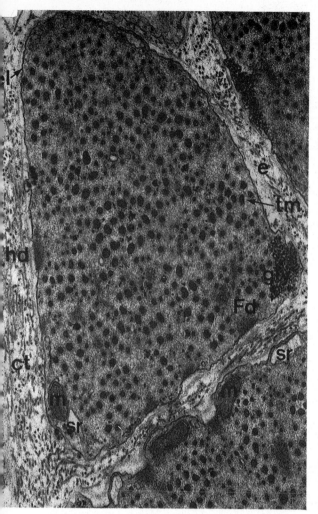

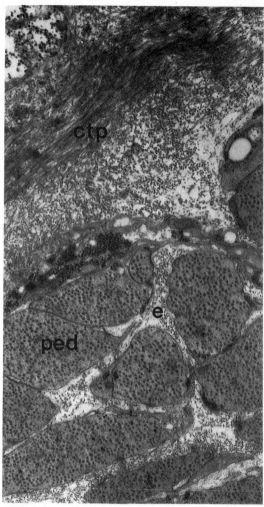

Above left

Fig. 126(c) – Cross-section through peduncle muscles of *P. lamarkii*. There is an irregular arrangement of thin and thick (tm) filaments, ill-defined electron-dense bodies (Fd) and peripheral mitochondria (m) present in the muscle fibrils. × 29 000 (Dr A. Bubel.)

Above right

Fig. 126(d) – Cross-section through the peduncle muscles (ped) of *P. lamarkii*. Thick collagenous connective sheets (ctp) (perimysium) surround the muscles, whereas thin collagenous connective tissue sheets (c) (endomysium) are found between them. × 4 700. (Dr A. Bubel.)

Right above

Fig. 127(a) – Longitudinal section through part of a smooth myoepithelial cell from brachiopod *(Terebratalia transversa)* tentacle. The thick myofilaments are extremely long, passing in and out of the plane of section. They are staggered throughout the myofilament field and are accompanied by numerous thin myofilaments. The circled thick filament is 88 nm in diameter. × 38 000. (Dr C.G. Reed.)

Right below

Fig. 127(b) – Transverse section through the smooth myoepithelial cell of a *T. transversa* tentacle. The contractile bundle consists of staggered thick paramyosinoid filaments surrounded by numerous thin filaments. × 21 300. Inset: an area of the myofilament field within the box. The thick filaments vary in diameter depending on the level they are sectioned at, and are surrounded by a profusion of thin filaments. × 64 400. (Dr C.G. Reed.)

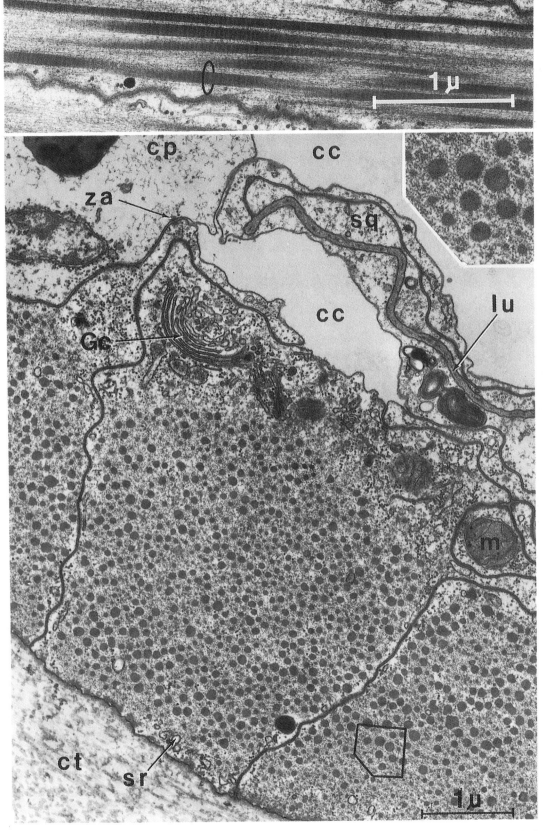

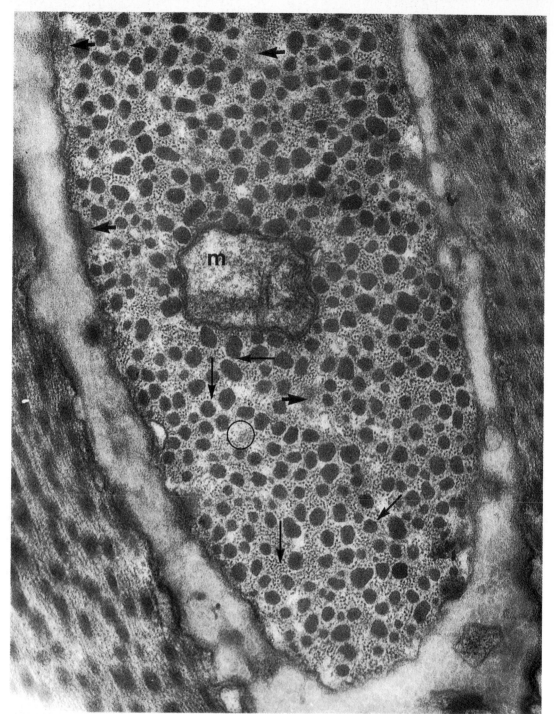

Fig. 128 – Transverse section of a smooth muscle from the gastropod *Buccinum undatum*. Groups of thin filaments (long arrows) show some degree of ordered packing as semi-circular orbits, isolated arcs, double rows and rosettes around thick filaments. Thick filaments cluster around a central mitochondrion (m). An electron-transparent zone free from thin filaments surrounds each thick filament. Dense areas apparently associated with the cell membrane as well as free in the cytoplasm (short arrows) may correspond to attachment plaques and dense bodies. Some of the free dense areas seem to be partially dissociated thick filaments (eg ringed area). × 61 000. (Dr S. Hunt, University of Lancaster.)

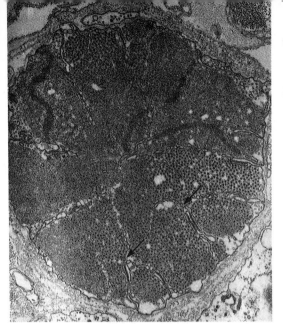

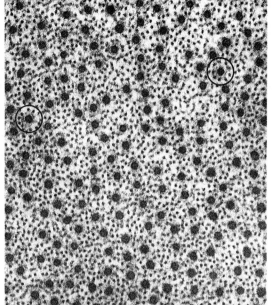

Fig. 129(a) – Transverse section through a slow abdominal receptor muscle of the crayfish, *Procambarius clarkii*. Myofilament bundles are not well demarcated by a membrane system, although the sacroplasm is radially compartimentalized by membrane structures. Diad and triads are also evident (arrows). × 13 500. (Dr T. Komuro.)

Fig. 129(b) – Transverse section through a slow muscle of *P. clarkii*, at the level of an A-band. Solid thick filaments are surrounded by more than six thin filaments. In places about 12 thin filaments surround some thick filaments (encircled). × 84 000. (Dr T. Komuro.)

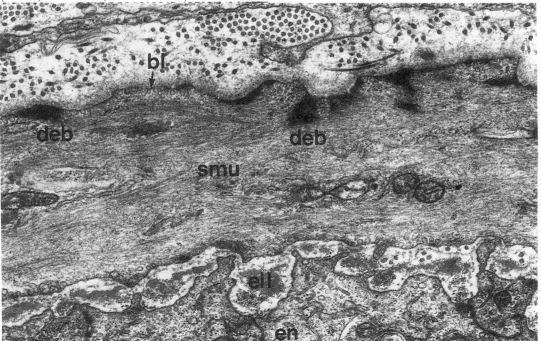

Fig. 130 – Arteriole (en) of the submucosa of the oesphagus of the rat. The distribution of dense bands (deb) is not uniform over the cell membranes of the smooth muscle cells (smu). They are markedly more numerous and conspicuous on the adventitial portion of the surface membrane than on its intimal portion. Bundles of actin penetrate into the dense bodies. × 20 000. (Dr G. Gabella.)
ell, elastic lamella.

thelium bordered by an outer basal lamina. However, although absent in most instances, a few sparsely distributed cells are found covering the internal surface of the luminal basal lamina (Figs 142, d, 143a, b, 144a, b). The latter have been variously interpreted as endothelial cells, amoebocytes and haemocytes. In some echinoderms, the blood vessel wall is composed simply of an endothelium of squamous epithelial cells, some bearing cilia, joined together by septate junctions (Fig 145a, b).

The presence of a basal lamina lining the lumina of some invertebrate blood vessels is in sharp contrast to the situation in most vertebrate blood vessels, where a basal lamina only borders the outer surface of the endothelium. The cells lining the outer face of the luminal basal lamina in annelids are regarded as peritoneal cells, and not true endothelial cells. Myofilaments are found in such cells and they have been variously referred to as myomesenthelial cells, myomesothelial cells, myoperithelial cells and myoepithelial cells. The contractile apparatus in these cells comprises both thick and thin filaments, and in terms of their dimensions, appearance and arrangement, they closely resemble thick paramyosin and thin actin filaments of invertebrate smooth muscle (Figs 141, 143, 144). The cells possess a prominent lobed cell body containing a nucleus and myofilament bundles, and extending from either side of the cell body are cytoplasmic processes which line the blood vessel lumen (Fig 141). The cytoplasmic processes become gradually protracted into tenuous cytoplasmic strands, on average 0.2 μm in width. Tenuous cytoplasmic strands of adjacent cells usually overlap one another. A few short junctions

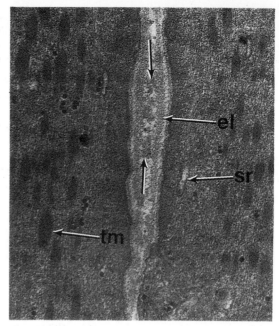

Fig. 131(b) – Cross-bridges (arrows) between the external laminae (el) of adjacent *P. lamarkii* peduncle muscles. × 66 000. (Dr A. Bubel.)

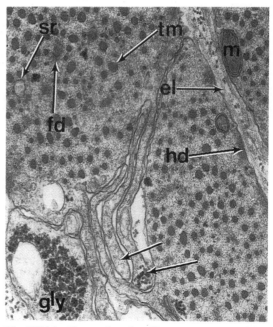

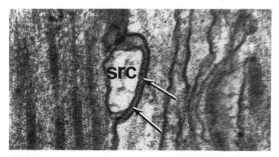

Fig. 131(a) – Subsacrolemmal cisternae (src) present along the peduncle muscle of *Pomatoceros lamarkii*. There is a specialization in the region where the cisternal membrane is closely apposed to the sacrolemma (arrows). × 80 000. (Dr A. Bubel.)

Fig. 131(c) – Peg-and-socket-like arrangement of the cytoplasmic processes (arrows) between the adjacent peduncle muscles of *P. lamarkii*. Among the thick myofilaments (tm) are ill-defined electron-dense free bodies (fd). × 28 000. (Dr A. Bubel.)

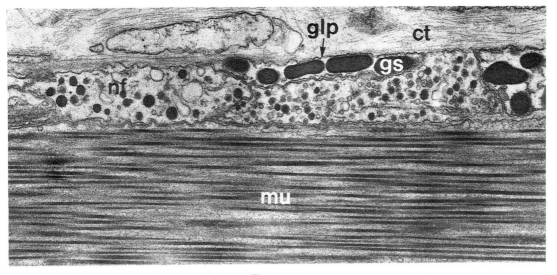

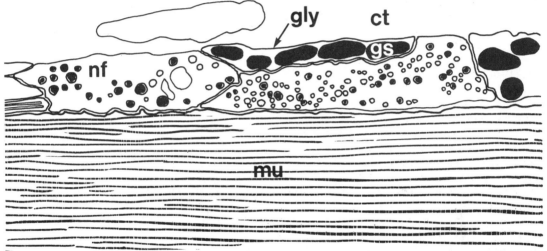

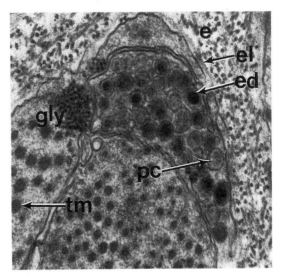

Top
Fig. 132(a) – Nerve fibres (nf) with glial processes (glp) running alongside the surface of an anterior byssus retractor muscle cell of the bivalve *Mytilus edulis*, forming myoneural junctions. A variety of vesicle types are present in the two nerve endings. × 16 000. (Dr O.C. McKenna.)

Centre
Fig. 132(b) – Diagram of Fig. 132(a).

Right
Fig. 133 – A nerve–muscle junction in *Pomatoceros lamarkii*. The nerve axon contains punctate (pc) and dense (ed) cored vesicles. × 60 000. (Dr A. Bubel.)

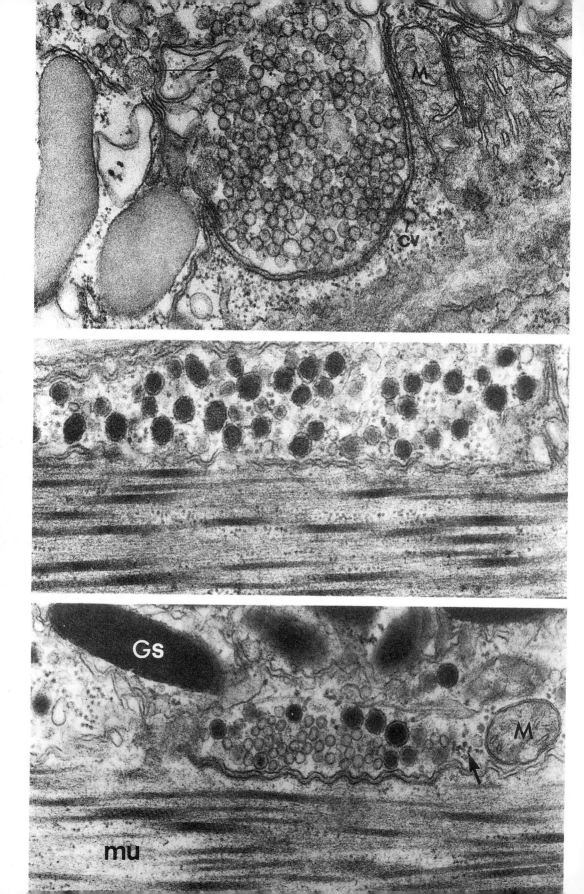

that resemble gap junctions appear to be responsible for connecting adjacent cells. The luminal surface of the blood vessel is generally smooth; however, a variable number of micro-villus-like protrusions project into the blood vessel lumen. In places, the blood vessel wall is discontinuous and gaps of between 0.5 and 6 μm in width are found, at which points the inner luminal and out basal laminae are contiguous (Fig 141c).

The luminal basal lamina not only provides structural support, but acts as a coarse filter. It acts as an effective partition between the blood vessel lumen and the surrounding tissue, confining blood-pigment molecules to the blood-vessel lumen. However, in localized regions, the luminal basal lamina may be permeable, allowing exchanges to take place between the blood and surrounding tissue. Both myoepithelial cells and coelomocytes in the blood vessel wall may, on occasion, endocytose blood-pigment molecules also (Fig 146). It has been established that in vertebrate blood vessels the basal lamina of endothelial cells is freely permeable to horseradish peroxidase (5 nm diameter, molecular weight 40 000) and to ferritin (10 nm diameter, molecular weight 462 000).

Blood cell types
In vertebrates, broadly three categories of blood cell are found, namely: red blood cells (erythrocytes, reticulocytes), white blood cells

Left top
Fig. 134(a) – Nerve ending in the anterior byssus retractor muscle (ABRM) of *Mytilus edulis* packed with clear vesicles. A few larger vesicles also contain punctate material (arrow). In the post-junctional cell a mitochondrion (m) is in close proximity to the muscle membrane and a coated vesicle (cv) is present. × 49 000. (Dr O.C. McKenna.)

Left centre
Fig. 134(b) – A nerve ending with extensive contact with an ABRM cell containing predominantly dense-cored vesicles. Only a few clear vesicles are present. × 61 000. (Dr O.C. McKenna.)

Left below
Fig. 134(c) – A nerve ending of ABRM containing a mixture of clear and dense cored vesicles, a mitochondrion (m) and some glycogen particles (arrow). Glial processes contain gliosomes (Gs). × 52 000. (Dr O.C. McKenna.)

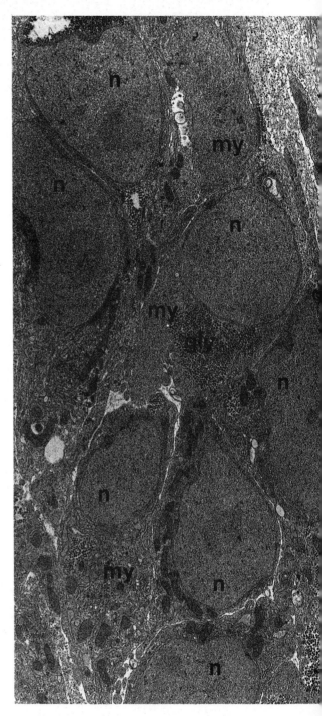

Above
Fig. 135 – Long strips of interdigitating cells, myoblasts (my), in the peduncle tissue of the polychaete, *Pomatoceros lamarkii*, two days after peduncle amputation. × 4 200. (Dr A. Bubel.)

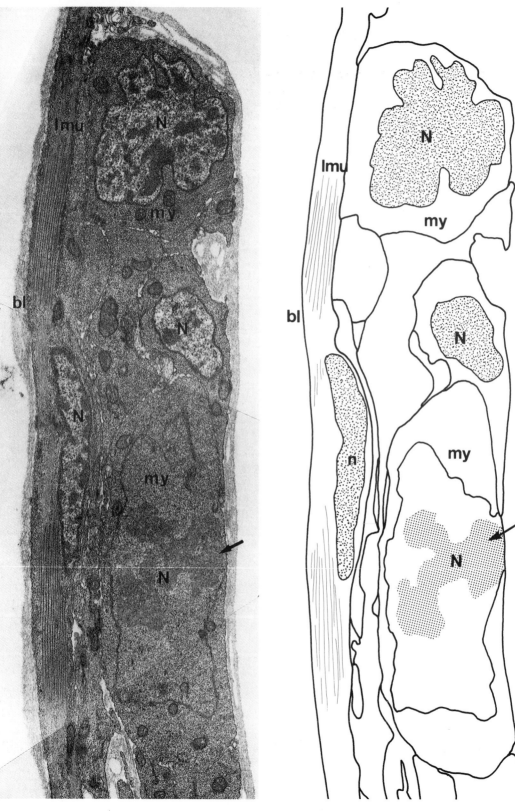

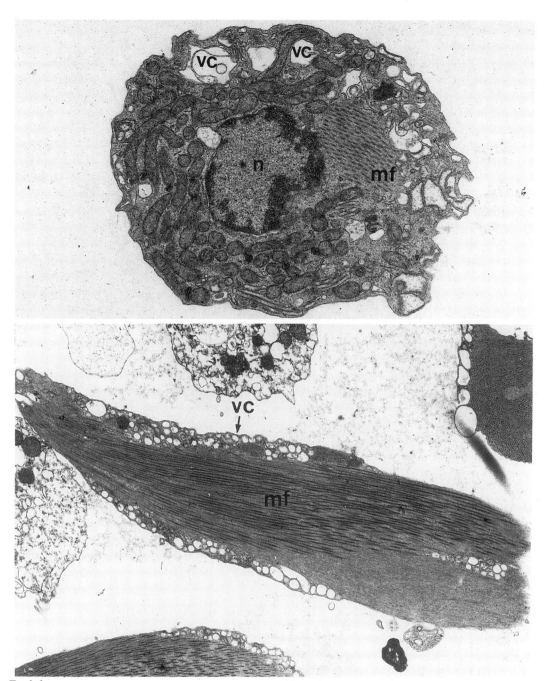

Far left

Fig. 136(a) – Longitudinal section of a portion of the dorso-longitudinal flight-muscle precursor of the insect *Chironomus* sp., at the beginning of the fourth (last) instar. Myoblasts (my) with dividing nucleus (arrow) are located beneath the muscle basal lamina (bl). × 15 000. (Dr M-C Lebart-Pedebas, University of Pierre and Marie Curie.)

Left

Fig. 136(b) – Diagram of Fig. 136(a).

Above top

Fig. 137(a) – Transverse section through the polar region of the spindle body cell of the brachiopod *Lingula anatina*. × 9 000. (Dr A. F. Rowley, University College of Swansea.)

Above

Fig. 137(b) – A spindle body cell of *L. anatina* with aligned fibres (mf) and peripheral/polar vacuolated region (vc). × 14 000. (Dr A.F. Rowley.)

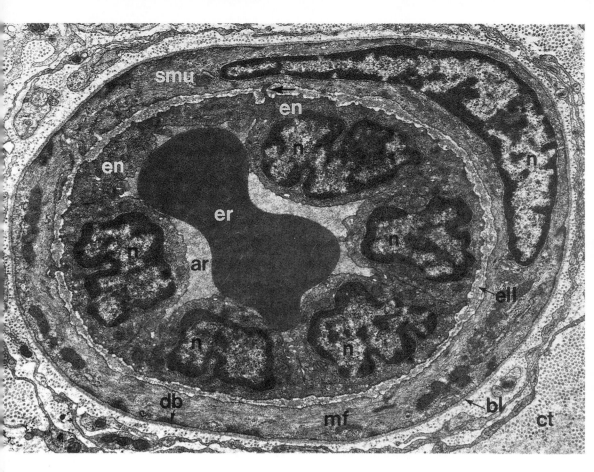

(granular leukocytes, lymphocytes, mono-cytes), and blood platelets. In invertebrates that possess a closed circulatory system, e.g. annelids, it is possible to differentiate between haemocytes of the circulatory system and coelomocytes of the coelom, or body cavity. However, in invertebrates that possess an open circulatory system, e.g. arthropods, the term haemocyte may include both coelomocytes and haemocytes. There is a multiplicity of cell types, the origin, nature and roles of which are often unclear. For example, there are found in some crustacea, haemocytes, hyaline cells (Fig 147a), granulocytes (Fig 147b), and intermedi-ate cells (Fig 147c); in some beetles, haemo-cytes, prohaemocytes (Fig 148a), plasmato-cytes (Fig 148b), cystocytes (Fig 148c), and granulocytes (Fig 148d); in some molluscs, haemocytes, small agranular basophils (Fig 149a, b), macrophages (Fig 150a, b), and acidophilic granulocytes (Figs 151a, b, c, 152a, b); in brachiopods, coelomocytes, erythro-cytes, phagocytic amoebocytes, and spindle body cells (Fig 137); in annelids, coelomocytes, agranular and granular ameobocytes (Fig 153a–c); in echinoderms, coelomocytes, pha-gocytic leukocytes (Fig 154a, b), lymphocytes (Fig 155), red spherule cells (Fig 156a), colour-less spherule cells (Fig 156b) and granulocytes (Fig 156c). Relatively undifferentiated precur-sor cells are found in some molluscs (Fig 151a); lymphocytes in echinoderms (Fig 155) and prohaematocytes in some arthropods (Fig 148a).

Several roles have been attributed to these cells, such as the transport of nutrients, carbo-hydrate metabolism, chitin synthesis, blood pigment synthesis, wound healing, haemo-lymph clotting, and phagocytosis (Fig 152a, b, 157). In the leaf beetle, *Ceratoma trifurcata*, granulocytes and plasmatocytes phagocytose plant viruses. In phagosomes, some of which occupy more than half of the cell volume, tobacco mosaic virus particles are tightly packed and form paracrystalline bodies com-posed of many layers of virus (Fig 158a, b).

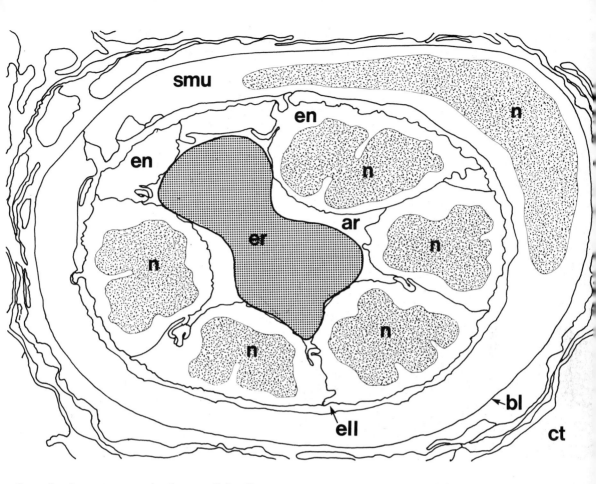

Granular haematocytes in the crayfish, *Orco-nectes virilis,* invade damaged tissue and form numerous microvillus-like processes which closely abut against both cellular and acellular components of damaged tissue and form junctional complexes (Fig 159). The damaged tissue is subsequently phagocytosed (Fig 160a, b).

Respiratory pigment synthesis

Blood pigments, haemoglobin, haemocyanin and chlorocruorin are found dissolved in the plasma of the circulatory systems of inverte-brates. Haemoglobin and chlorocruorin mole-cules appear to be synthesized in cells lying upon the outer coelomic surface of blood vessels in annelids. Extravasal cells in some polychaetes are responsible for chlorocruorin molecule synthesis, whereas in some oligo-chaetes, a plug of tissue 'the heart body' synthesizes haemoglobin molecules. The cells of both annelids possess an extensive granular ER and well developed Golgi complexes. The

Left
Fig. 138(a) – Transverse section of an arteriole in the submucosa of the oesophagus of a rat. The tunica media shows a single smooth muscle cell (smu) which contains filaments (mf) and dense bands (db) which are found almost exclusively on the membranes of the adventitial surface of the cell. The intimal aspect of the muscle cell is associated with an inner elastic lamella (ell). Finger-like processes (arrows) of the endothelial cells (en) pierce the elastic lamella and are closely apposed to the muscle cell membrane or invaginate it. The adventitial aspect of the muscle cell is covered by a basal lamina (bl). × 25 000. (Dr G. Gabella.)

Above
Fig. 138(b) – Diagram of Fig. 138(a).

blood pigment molecules appear to be synthesized by these two organelles (Figs 161a, b, 162a, b). Secretory vesicles containing the molecules are believed to cross the vascular lamina to reach the blood vessel lumen. However, in some arthropods, cyanoblasts are responsible for the synthesis of haemocyanin molecules. In the cytoplasm free ribosomes are implicated in the synthesis of the molecules. The haemocyanin molecules accumulate in the cytoplasm in numerous crystalline arrays. These are composed of hexagonally packed hollow-appearing cylinders with a diameter of about 19 nm and a centre-to-centre spacing of 26 nm (Fig 163a, b). Longitudinal stacking of the molecules composing the cylinders is evident in a 10 nm transverse periodicity (Fig 164). Numerous small crystals subsequently grow to form a large crystal, which may occupy most of the cell volume of the cyanocyte (Fig 34). The cell then ruptures and liberates the haemocyanin crystal into the haemocoel. In molluscs, specialized pore cells in the connective tissue are responsible for the synthesis of haemocyanin crystals in some freshwater and terrestrial gastropods and for haemoglobin in other freshwater gastropods. The pigment molecules are synthesized in granular ER cisternae and released by exocytosis (Fig 33). The Golgi apparatus is by-passed in this secretion process.

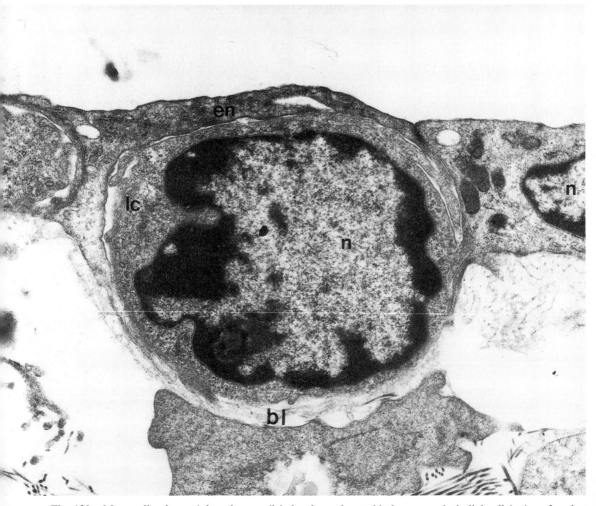

Fig. 139 – Mammalian lung. A lymphocyte (lc) that has migrated below an endothelial cell (en) surface but is still above the endothelial basal lamina (bl). The endothelial layer above the lymphocyte is intact. × 14 000. (Dr E.A. Perkett, Vanderbilt University.)

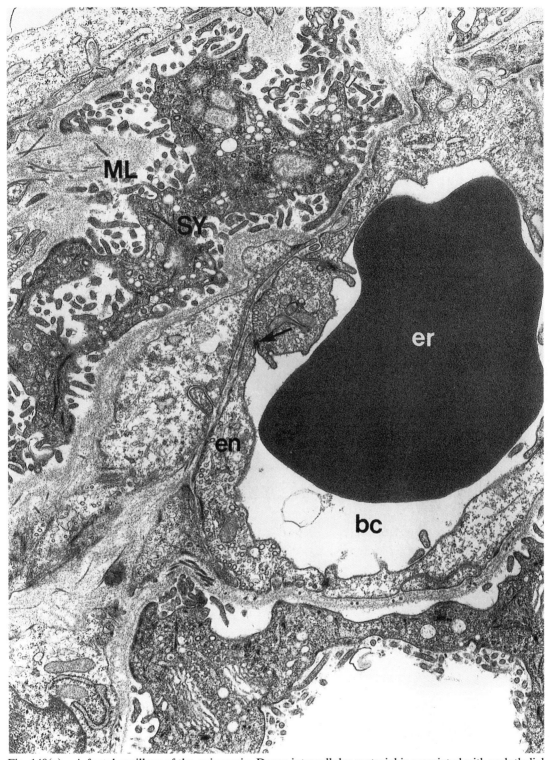

Fig. 140(a) – A foetal capillary of the guinea pig. Dense intracellular material is associated with endothelial cell (en) zonular junctions (arrow). × 11 500. (Dr J.A. Firth, St. George's Hospital Medical School.) ML, maternal lacuna; SY, syncytotrophoblast.

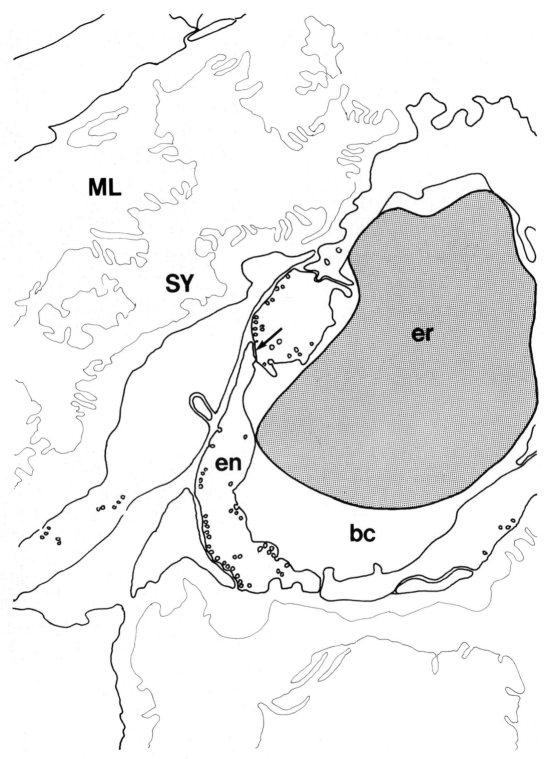

Fig. 140(b) – Diagram of Fig. 140(a).

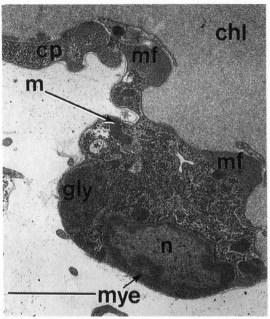

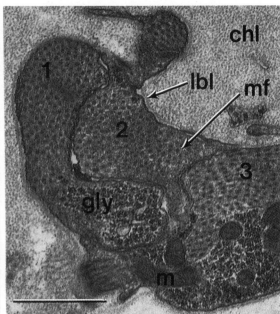

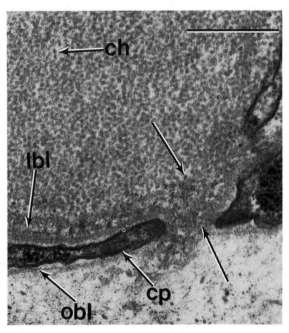

Above left
Fig 141(a) – Longitudinal section through an opercular blood vessel of the polychaete, *Pomatoceros
lamarkii*. The blood vessel wall is made up of myoepithelial cells. Myoepithelial cells possess a lobed
cell body containing a nucleus with laterally extended cytoplasmic process (cp) containing myofilaments (mf). × 12 000. (Dr A. Bubel.)
chl, chlorocruorin molecules.

Above right
Fig. 141(b) – The interdigitation of three myoepithelial cell lateral processes (numbered) with well
endowed with myofilaments (mf) in the blood vessel
wall of *P. lamarkii*. × 12 500. (Dr A. Bubel.)

Right
Fig. 141(c) – A gap in the blood vessel wall of *P.
lamarkii* where luminal (lbl) and outer (obl) basal
laminae are contiguous (arrows). In places the blood
vessel wall is lined by tenuous cytoplasmic strands
(cp) extending from adjacent myoepithelial cells.
× 12 500. (Dr. A. Bubel.)

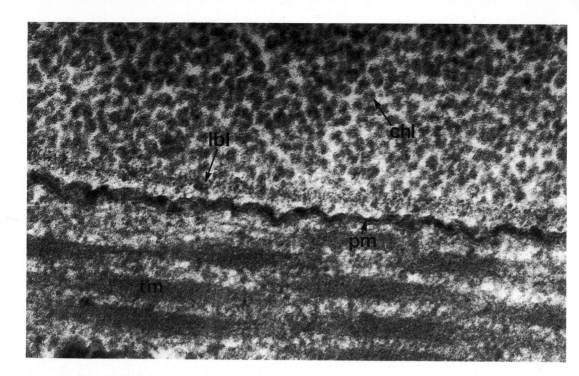

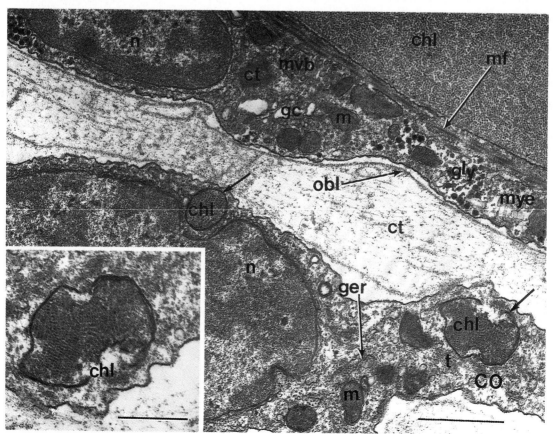

GAMETOGENESIS

Gametogenesis refers to the production of gametes in the germ tissue of animals (the similar process in plants is called sporogenesis). In the male, gametogenesis is called spermatogenesis, and in the female, oogenesis.

Spermatogenesis

The production of mature spermatozoa in the testes involves cellular divisions and maturations. This entire process is called spermatogenesis. In the testes, the germ cells, spermatogonia, give rise by mitotic proliferation to primary spermatocytes. The primary spermatocytes go through meiosis (reduction division) that involves two divisions. During the first meiotic division, the primary spermatocytes give rise to secondary spermatocytes during which time there occurs 50% reduction in the number of chromosomes. The secondary spermatocytes thus formed rapidly divide by regular mitosis to give rise to spermatids (Fig 165). This second meiotic division does not result in a further reduction in chromosome number. The spermatids do not further divide, but go through a cytological transformation that leads to mature spermatazoa.

Testes are composed of germinal and secretory regions. The epithelial cells of the secretory region are capable of phagocytosing mature spermatozoa, in addition to secreting protein substances into the lumen of the testes (Fig 165a). In some species, the germinal region comprises somatic cells surrounding cysts of germ cells (Fig 165c). Spermatogonia tend to have a dense nucleus, with heavily staining chromatin with prominent nucleoli and a dense cytoplasm rich in free ribosomes

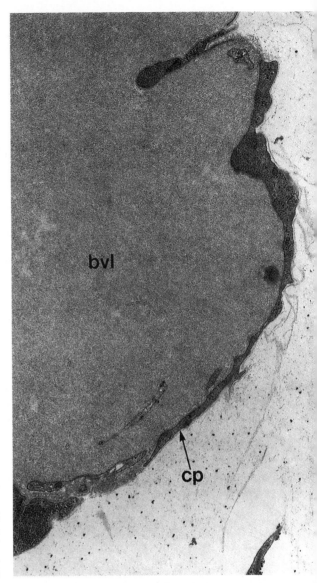

Left above
Fig. 142(a) – High-magnification electron micrograph of the luminal basal lamina (lbl) of the opercular blood vessel of *Pomatoceros lamarkii.* × 100 00. (Dr A. Bubel.)

Left below
Fig. 142(b) – A coelomocyte-like (CO) cell in the connective tissue (ct) of the operculum of *P. lamarkii,* closely associated with a blood vessel myoepithelial cell (mye). Membrane-bound bodies (arrows) in the coelomocyte cell contains chains of chlorocruorin molecules (chl). × 24 000. Inset: a chlorocruorin-containing body. × 38 000. (Dr A. Bubel.)

Above
Fig 142(c) – Longitudinal section through an opercular blood vessel of *P. lamarkii.* × 12 000. (Dr A. Bubel.)

bvl, blood vessel lumen; cp, cytoplasmic extensions of myoepithelial.

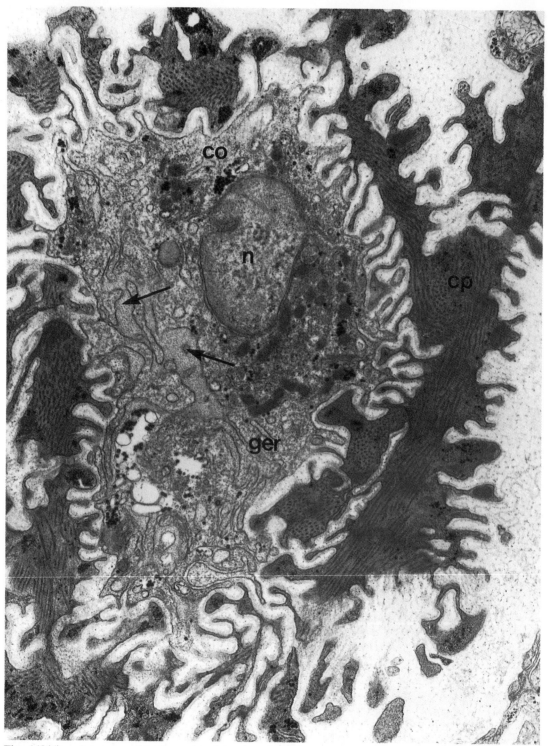

Fig. 142(d) – A cross-section through a lateral outpushing of the opercular blood vessel containing an irregularly outlined coelomocyte-like cell (co) with irregularly branched cisternae of the endoplasmic reticulum, distended by chlorocruorin molecules (arrows). × 23 000. (Dr A. Bubel.)
cp, myoepithelial cell process containing myofilaments.

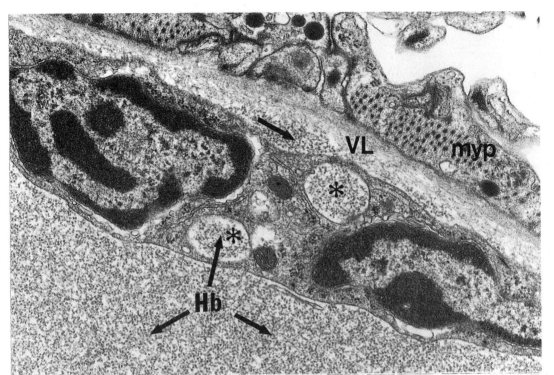

Fig. 143(a) – Longitudinal section through a follicular-supply blood vessel of the annelid *Amynthas sp*. The blood vessel wall is composed of myoepithelial cells (myp) with a vascular lamina (VL) containing collagen fibrils. An elongate littoral cell with two haemoglobin (Hb) pinosomes (asterisk) is present in the blood vessel lumen. Haemoglobin (upper arrow) molecules are interspersed between the littoral cell and the vascular lamina. × 18 000. (Dr M.M. Friedman, University of Pennsylvania.)

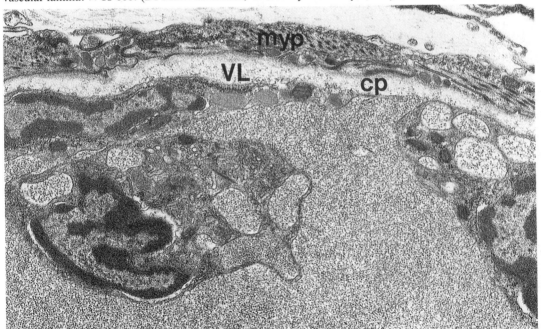

Fig. 143(b) – Ordered thick and thin filaments in myoepithelial cells (myp) of *Amynthas* blood vessel. × 69 000. (Dr M.M. Friedman.)   VL, vascular lamina; cp, radiate cell process.

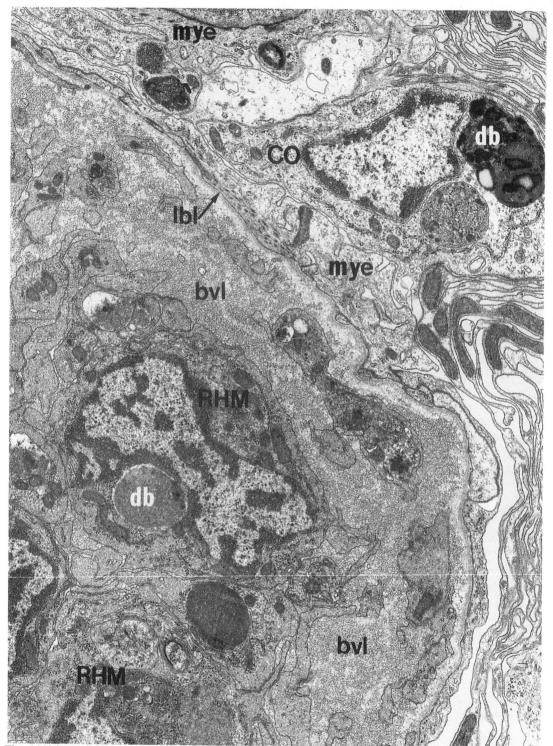

Fig. 144(a) – A portion of a nephridial capillary dilation of the annelid, *Lumbricus,* containing radiate haemocytes (RHM). Electron-dense bodies (db) are closely associated with the nucleus of radiate haemocytes and are also found in myoepithelial cells (mye) and in overlying colemocytes (CO). × 16 000. (Dr M.M. Friedman.)

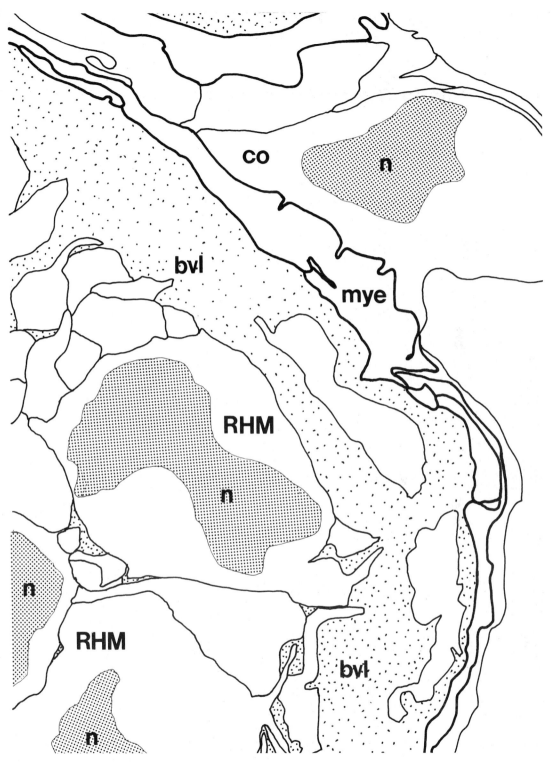

Fig. 144(b) – Diagram of Fig. 144(a).

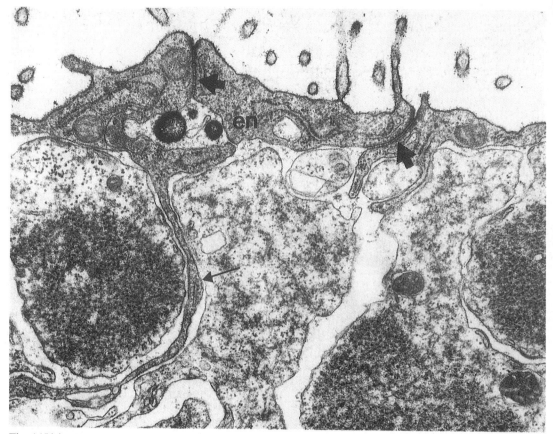

Fig. 145(a) – Endothelial cells (en) of the sea urchin *Strongylocentrotus franciscanus* joined by apical septate junctions (thick arrows). A cell process (thin arrow) extends towards the basal lamina bordering the connective tissue layer to form a foot-like plate. × 18 000. (Prof. Dr E. Florey.)

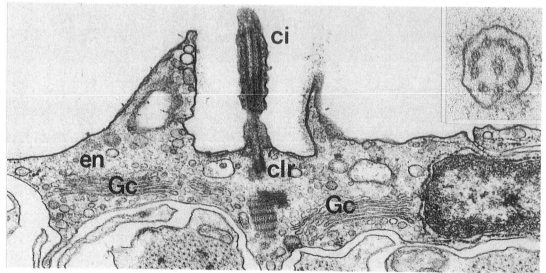

Fig. 145(b) – Part of an endothelial cell (en) from the sea urchin *Echinus esculentus* containing a portion of a cilium (ci) and ciliary rootlet (clr) flanked by a well-developed Golgi complex (Gc). × 29 000. Inset: cross-section of a cilium of *E. esculentus* with a typical 9+2 structure. × 56 000. (Prof. Dr E. Florey.)

224

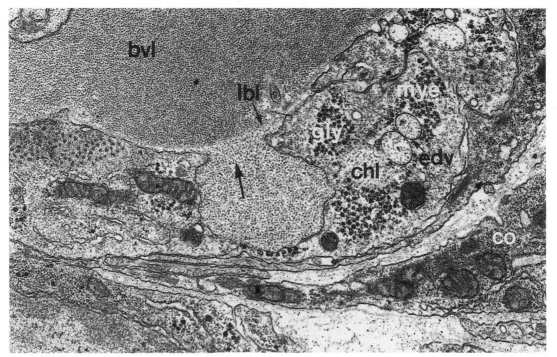

Fig. 146(a) – The peduncle blood vessel of the polychaete *Pomatoceros lamarkii* 24 hours after opercular filament amputation. Chlorocruorin molecules appear to be endocytosed (arrow) from the blood vessel lumen (bvl), into endocytotic vesicles (edv) in the cytoplasm of a myoepithelial cell. There are also free chlorocruorin molecules (chl) in the cell cytoplasm. × 18 900. (Dr A. Bubel.)

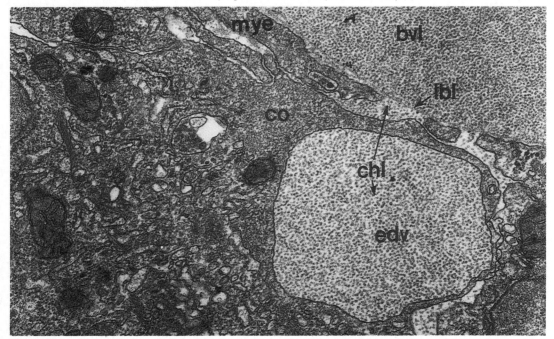

Fig. 146(b) – A coelomocyte (co) in the blood vessel wall of the polychaete, *P. lamarckii* containing a large endocytotic vacuole (edv) filled with chlorocruorin molecules (chl). The intercellular space between the coelomocyte (co) and the myoepithelial cell (mye) is filled with chlorocruorin molecules. × 28 000. (Dr A. Bubel.)

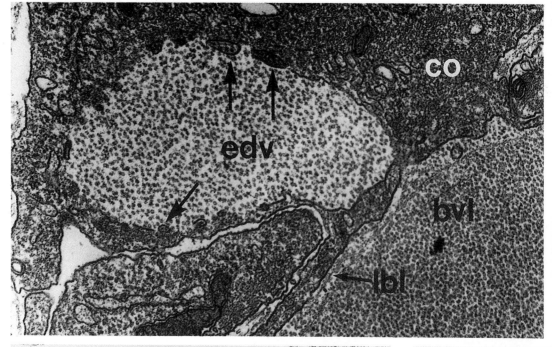

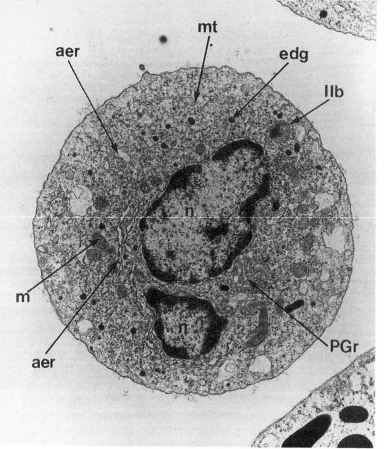

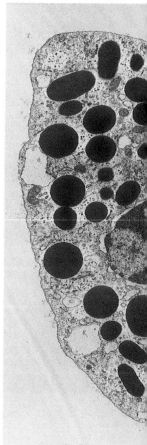

Left

Fig. 146(c) – A fragmented bordering membrane of an endocytotic vacuole (edv) in a coelomocyte (Co) appears as a series of small peripheral vesicles which contain chlorocruorin molecules (arrows). × 42 000. (Dr A. Bubel.)

Below left

Fig. 147(a) – A hyaline haemocyte of the crab *Callinectes sapidus*. × 9 710. (Dr J.E. Bodammer, National Marine Fisheries Service, Maryland.)
edg, electron-dense granules;
llb, lysosomal-like body; PGr, presumptive Golgi region.

Below centre

Fig. 147(b) – A granulocyte from the crab *C. sapidus*. × 6 750. (Dr J.E. Bodammer.)
elb, electron-lucent inclusion body;

Below right

Fig. 147(c) – An intermediate cell from the crab, *C. sapidus*. × 6 500. (Dr J.E. Bodammer.)
stg, 'striated' granule composed of orderly arranged filaments.

(Fig 166). Gradually, in spermatocytes, the cytoplasm enlarges and the chromatin pattern of the nuclei is rearranged and shows thick filamentous chromosomes (Fig 167). The cytoplasm of spermatocytes is small in relation to the nucleus and contains few mitochondria and many free ribosomes. In early spermatids, two centrioles and a flagellum rudiment are present near the nucleus. The shape of the nucleus changes in some species from a spherical shape in early stages of differentiation to an elongate and barrel shape. The nucleus also changes from a homogenous moderately electron-dense appearance to a more condensed appearance (Fig 168a). During this time, mitochondria, which are randomly distributed, increase in size but decrease in number, and accumulate at one pole, close to the insertion of the flagellum. Golgi complexes in the cells appear to play a role in the formation of acrosomal granules, which settle on the top pole of the nucleus to form the acrosomal cap. The shape and structure of the acrosomal cap differs in

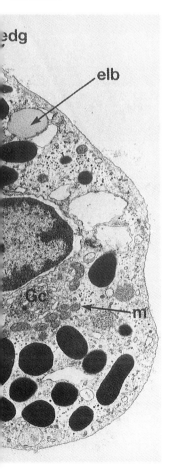

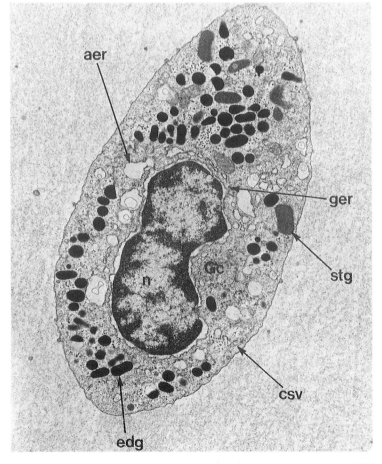

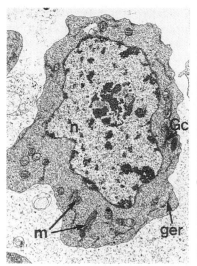

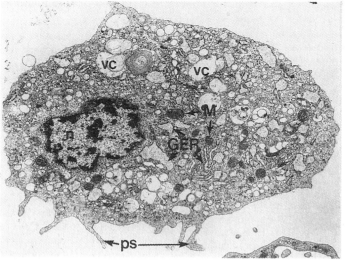

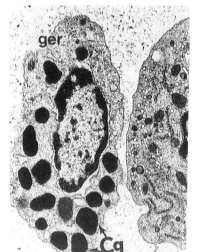

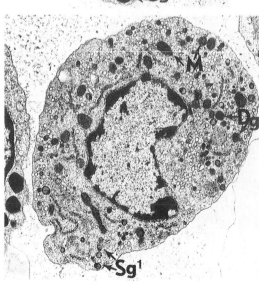

Above left
Fig. 148(a) – A typical prohaemocyte, from the beetle *Ceratoma trifurcata*, with a large nucleus (n), a Golgi complex (GC), small number of mitochondria (m), and a few segments of granular endoplasmic reticulum (ger) present in a free-ribosome-rich ground cytoplasm. × 5 700. (Dr K.S. Kim.)

Above right
Fig. 148(b) – A plasmatocyte from the beetle *C. trifurcata*, containing well-developed cytoplasmic organelles and vacuoles (vc). A few psuedopodia (ps) are formed at the cell surface. × 5 700. (Dr K.S. Kim.)

Centre left
Fig. 148(c) – An atypical cystocyte, from the bean leaf beetle, *Ceratoma trifurcata*, with cytoplasm filled with numerous large homogeneously electron-dense granules (Cg). A few segments of granular endoplasmic reticulum are also present. × 6 000. (Dr K.S. Kim, University of Arkansas.)

Below left
Fig. 148(d) – A granulocyte from the beetle *C. trifurcata*, containing characteristic electron-dense granules (Dg) throughout the cytoplasm. × 6 000. (Dr K.S. Kim.)   Sg, structured granules.

Right above
Fig. 149(a) – A small basophil cell from the bivalve *Mytilus edulis*. × 30 000. (Dr A. Bubel.)

Right below
Fig. 149(b) – An aggregation of small basophil cells (LYM) from the bivalve, *M. edulis* containing a well-developed granular endoplasmic reticulum (GER) (arrows) and vesicles (V) in their cytoplasm. × 12 500. (Dr A. Bubel.)   DGR, developing granulocyte.

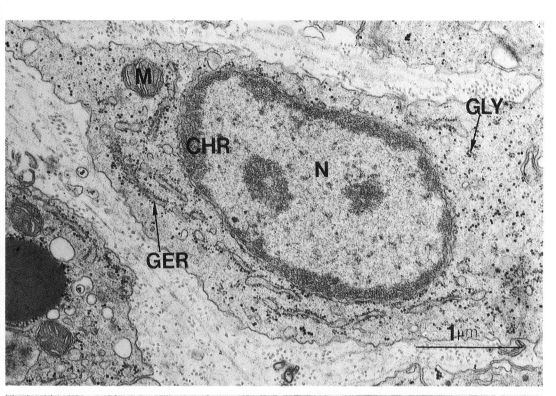

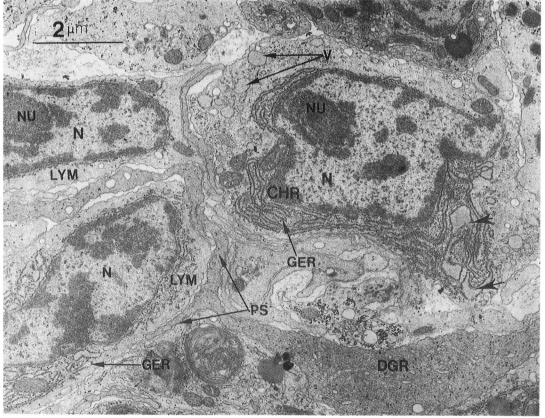

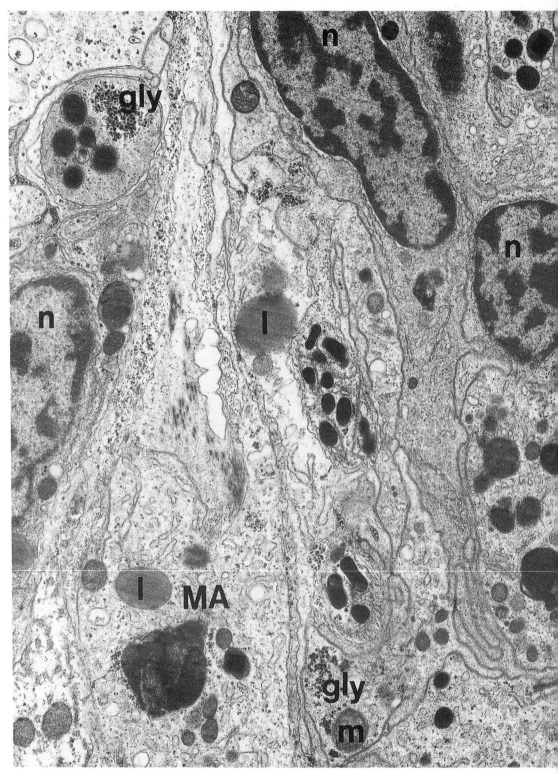

Fig. 150(a) – Macrophages (MA) with lysosomes (ly) and secondary lysosomes (sly) in the mantle tissue of the bivalve, *Mytilus edulis*. × 12 500. (Dr A. Bubel.)

230

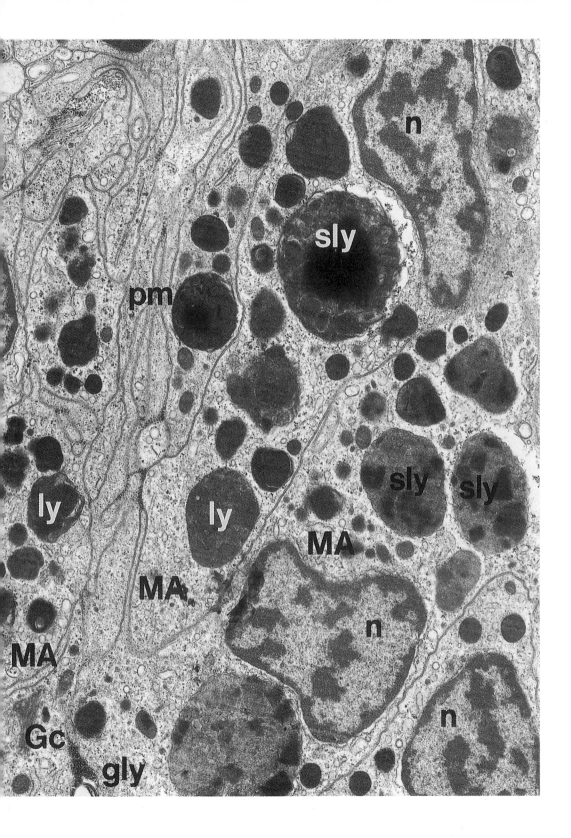

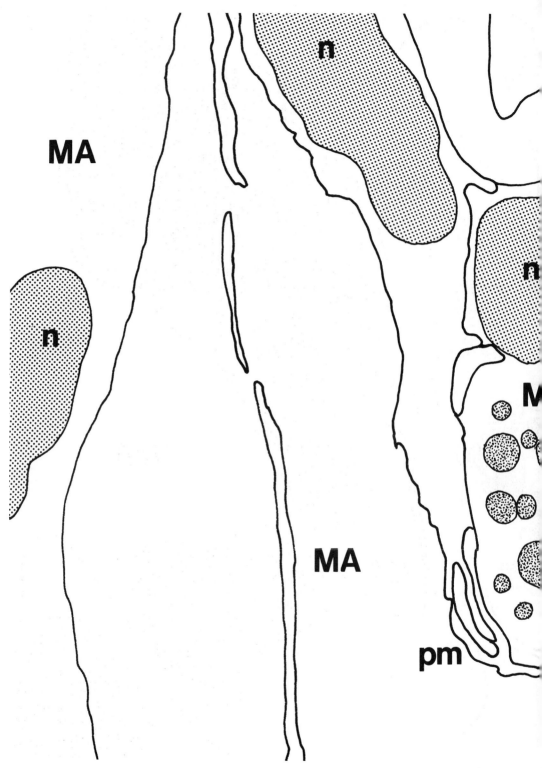

Fig. 150(b) – Diagram of Fig. 150(a).

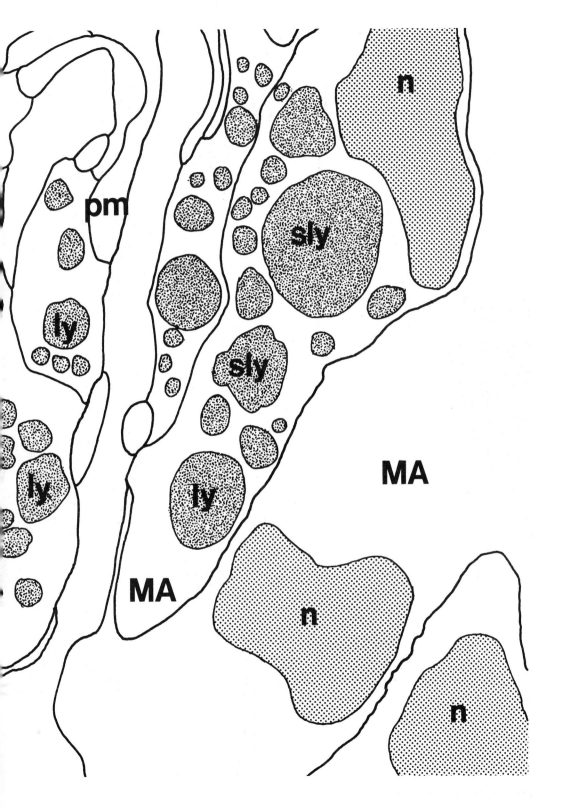

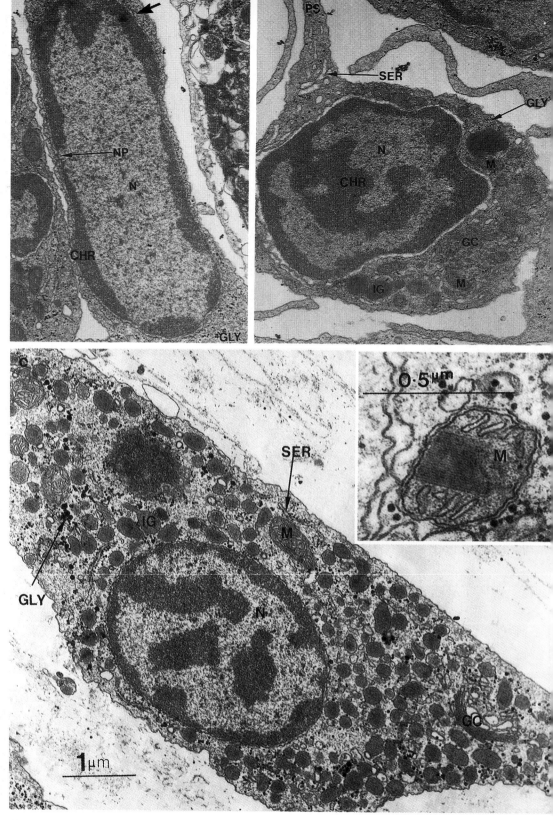

Far left
Fig. 151(a) – A granulocyte precursor cell of *Mytilus edulis* with a noticeable absence of organelles in the narrow strip of cytoplasm (arrow). × 10 500. (Dr A. Bubel.)

Left
Fig. 151(b) – A developing granulocyte of *Mytilus edulis*. × 11 000. (Dr A. Bubel.)
IG, inclusion granule.

Below left
Fig. 151(c) – A granulocyte of *Mytilus edulis* in a further stage of development. × 12 000.
Inset: A mitochondrion (m) containing a crystalline-like lattice. × 80 000. (Dr A. Bubel.)

Right
Fig. 152(a) – A granulocyte of *Mytilus edulis* containing phagolysomes (PHL) in the peripheral cytoplasm. × 7 000. (Dr A. Bubel.)

Below
Fig. 152(b) – A granulocyte of *Mytilus edulis* showing a greater development of phagolysosomes (PHL) in the cytoplasm. × 17 000. (Dr A. Bubel.)

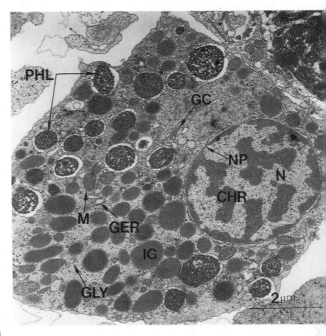

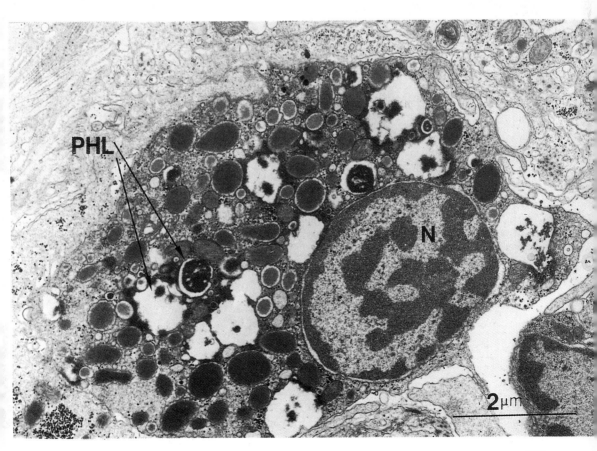

235

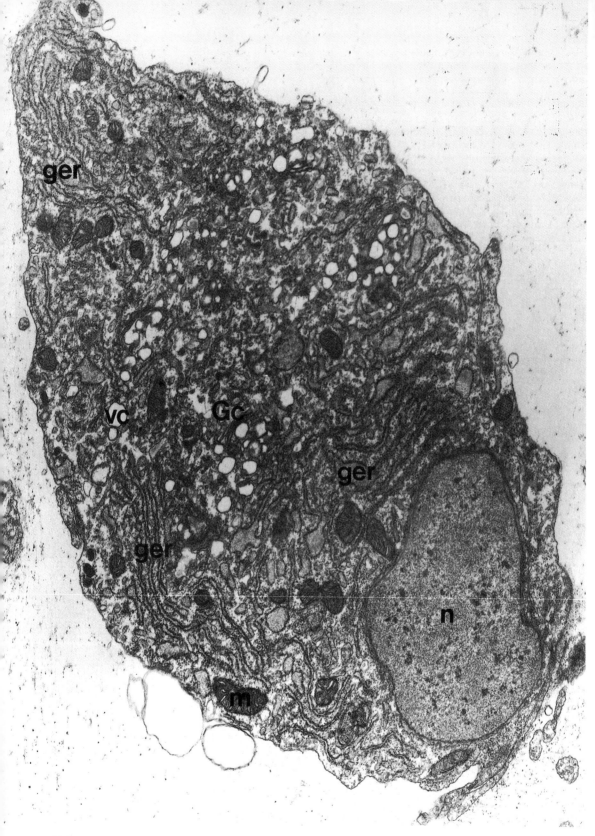

Left
Fig. 153(a) – The features of a coelomocyte-like cell from the connective tissue of the peduncle of the polychaete, *Pomatoceros*. × 6 700. (Dr A. Bubel.)

Right
Fig. 153(b) – Type 11 amoebocyte with characteristic pleomorphic nucleus (N) and prominent nucleolus (NU) from the polychaete *Nicolea zostericola*. × 4 000. (Dr K.J. Eckelbarger.)

Below
Fig. 153(c) – Nucleolar lobe (N) of type 11 amoebocyte with prominent nucleolus (NU) from polychaete *N. zostericola*. × 23 000. (Dr K.J. Eckelbarger.)

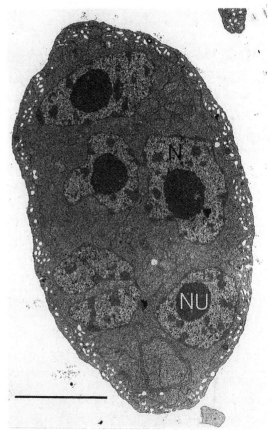

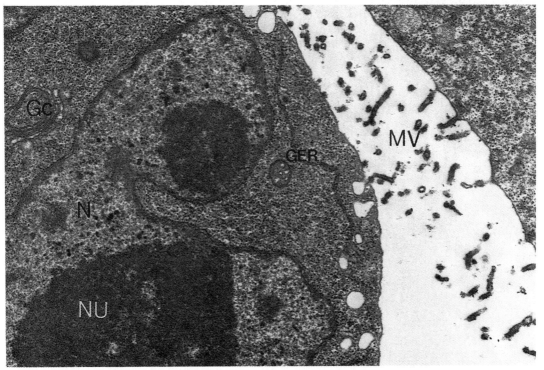

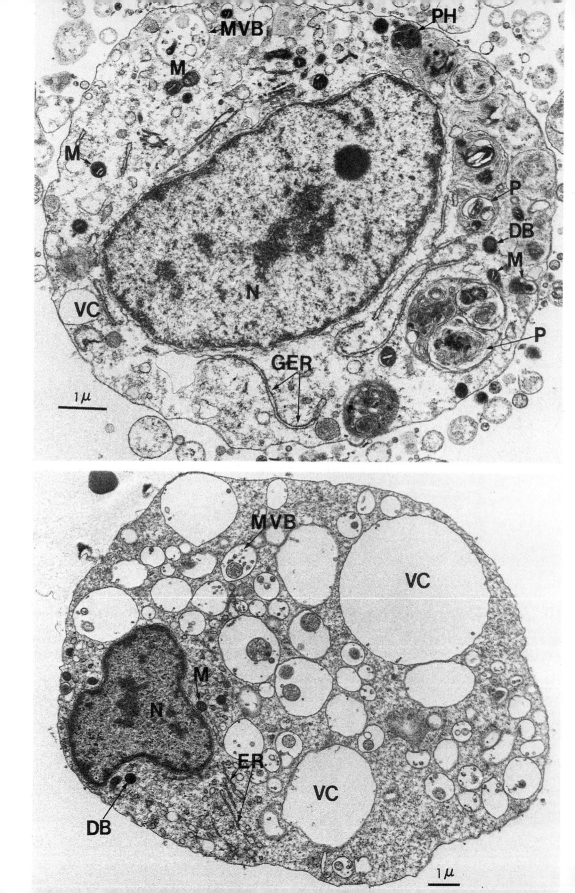

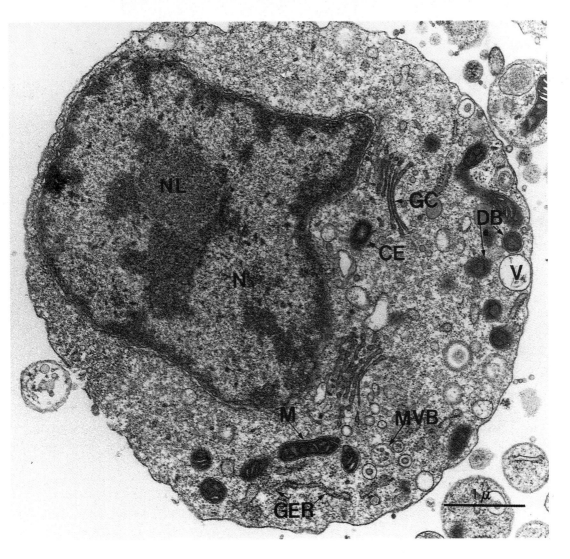

Left above
Fig. 154(a) – Phagocytic leukocyte from the sea urchin, *Stronglyocentrotus drobachiensis*. × 13 000. (Dr V.G. Vethamany, Dalhousie University.) DB, small electron-dense body.

Left below
Fig. 154(b) – Phagocytic leukocyte with vacuoles (Vc) from *S. drobachiensis*. × 7 000. (Dr V.G. Vethamany.) DB, small electron-dense body.

Above
Fig. 155 – Lymphocyte from sea urchin *Strongylocentrotus drobachiensis*. × 23 000. (Dr V.G. Vethamany.) DB, small dense body.

different species. It becomes the front of the spermatozoon (Fig 168a–c). The mature spermatozoon consists of structurally distinct parts, namely a head – contains the nucleus – a neck – contains a pair of centrioles – and a tail made up of middle, principal and end pieces, usually composed of an axoneme with a 9+2 structure (Fig 168a). The plasma membrane in mammalian spermatozoa is also partitioned into anatomically and biochemically distinct regions (Fig 168b, c). In some invertebrate spermatozoa, the cytoplasm is ensheathed by two electron-dense plates (Fig 165c, d).

Although spermatozoa are formed by conventional spermatogenesis, sperm pairing takes place in some animals. Spermatozoa, in some species are joined in pairs by septate

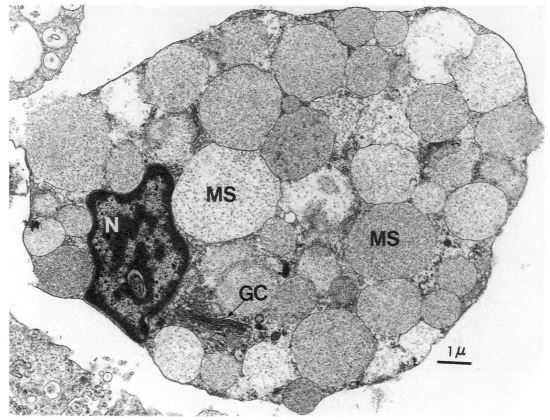

Fig. 156(a) – Red spherule cell from sea urchin, *Strongylocentrotus drobachiensis*. × 9 000. (Dr V.G. Vethamany.)
MS, membrane-limited spherules.

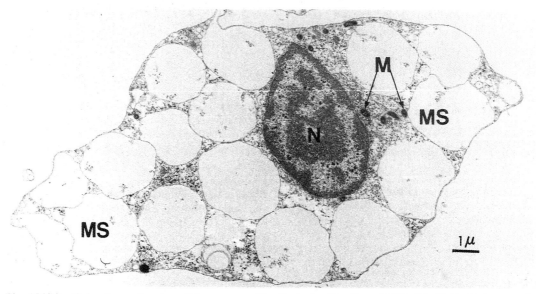

Fig. 156(b) – Colourless spherule cell from sea urchin, *S. drobachiensis*. × 7 000. (Dr V.G. Vethamany.)
MS, 'colourless' membrane-bound spherules.

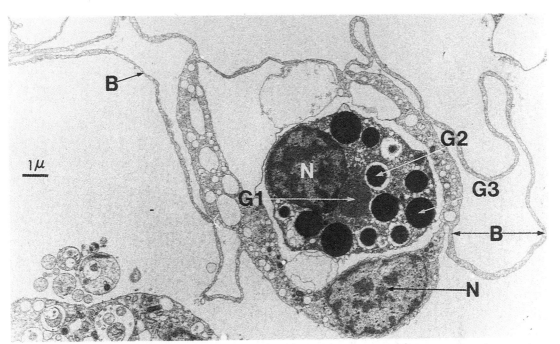

Fig. 156(c) A phagocytic leukocyte from *S. drobachiensis*, with veils of bladder-like cytoplasmic extensions surrounding a granulocyte. × 7 000. (Dr V.G. Vethamany.)
B, bladder-like cytoplasmic extension of the phagocyte; G1, G2, G3, morphologically different granules in granulocyte.

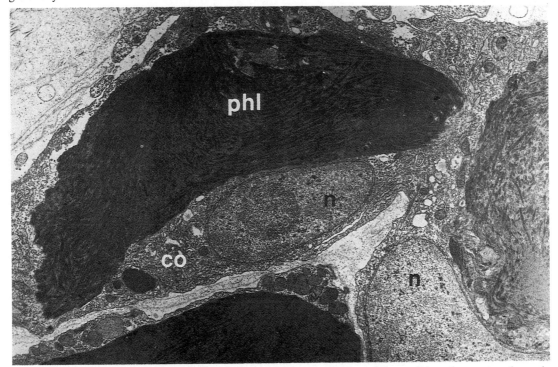

Fig. 157(a) – A coelomocyte (co) from polychaete, *Pomatoceros lamarkii* containing phagocytosed muscle fragments. × 15 000. (Dr A. Bubel.)
phl, phagolysosome containing muscle fragments.

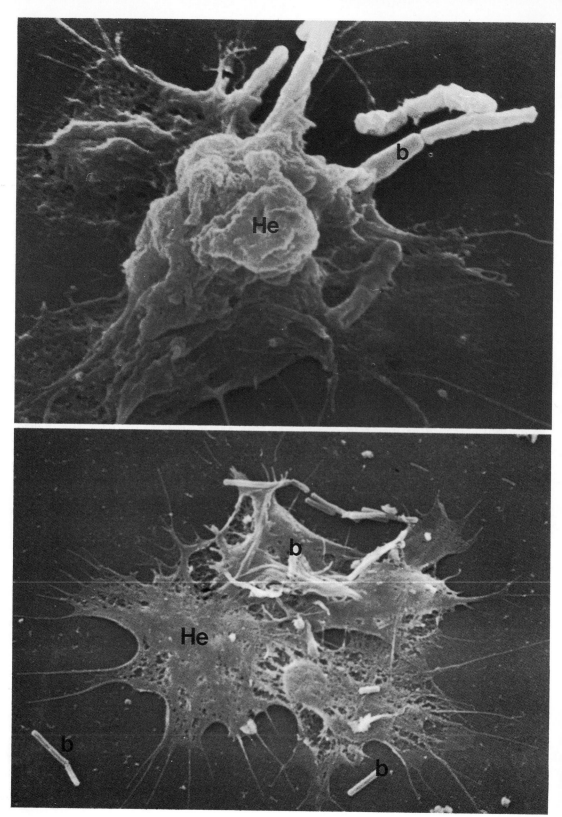

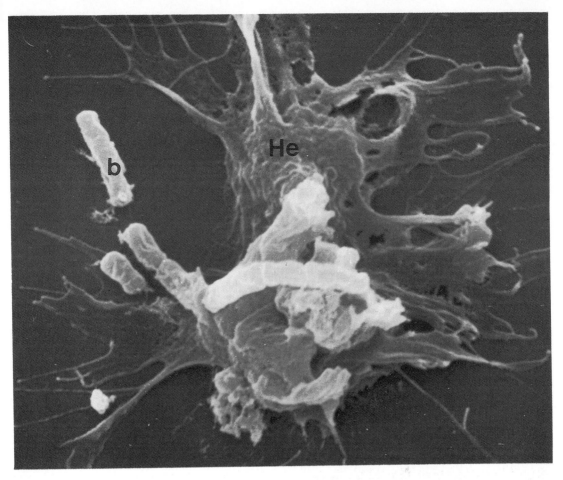

junctions. Also, atypical spermatozoa may arise; for example, in the gastropod *Theodoxus fluviatilis,* the final transformation of a spermatid into a spermatozoa is effected by the expulsion of the nucleus from the cytoplasm (Fig 169a, b). Each spermatozoon thus produced is a monoflagellate cell completely lacking a nucleus. At least 12 μm of the axoneme of the long flagellum in the polychaete *Questa* spermatozoa is invested by two mitochondrial derivatives, each with a narrow crescentric transverse profile, posteriorly limited by an annulus. The axoneme is of the 9+2 type, but in the mitochondrial region as many as 27 accessory peripheral microtubules in nine triads encircle it. At each of the nine intervals between adjacent doublets is a triad of three microtubules, which are clearly continuous with the centriolar triads (Fig 170a). More posteriorly in the mitochondrial region, the number of accessory microtubules is reduced (Fig 170b, c). Ultimately, only nine accessory

Left above
Fig. 157(b) – Haemocyte (He) of gastropod *Patella vulgata* entrapping/phagocytosing the bacterium *Bacillus subtilis,* (b) added to haemolymph. × 10 900. (A. Hoskins, Portsmouth Polytechnic.)

Left below
Fig. 157(c) – Haemocyte of *P. vulgata* entrapping/phagocytosing *B. subtilus* (b). × 11 400. (G. Johnstone, Portsmouth Polytechnic.)

Above
Fig. 157(d) – Haemocyte of *P. vulgata* entrapping/phagocytosing *B. subtilus* (b). × 17 150. (C. Bartram, Portsmouth Polytechnic.)

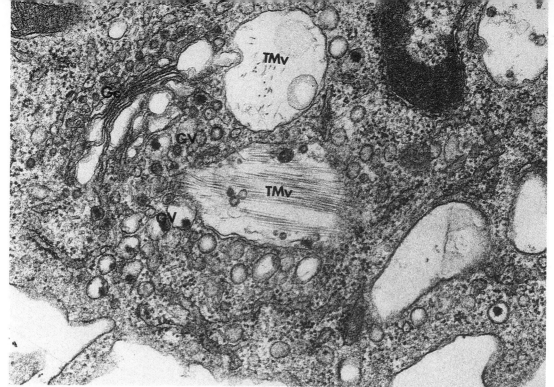

Fig. 158(a) – Granulocyte of the bean leaf beetle, *Ceratoma trifurcata* Golgi-derived vesicles (GV) containing electron-dense material are closely associated with tobacco mosaic virus (TMV) containing phagosomes (TMV). Also inside the phagosome are a few vesicles making contact with TMV particles. × 28 000. (Dr K.S. Kim.)

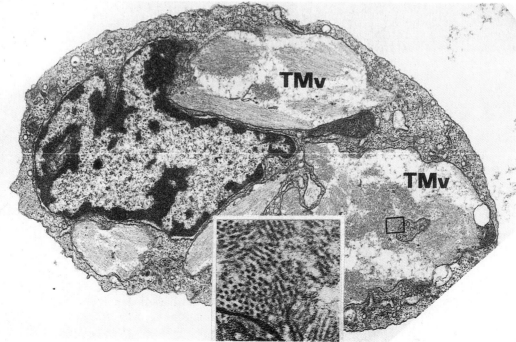

Fig. 158(b) – Plasmatocyte from *C. trifurcata* with large TMV-containing phagosomes (TMv) in the cytoplasm. TMV-free areas between TMV bundles or in the periphery of phagosomes are also evident. × 15 000. Inset: tightly packed and structurally unaltered TMV particles in phagosome lumen. × 95 000. (Dr K.S. Kim.)

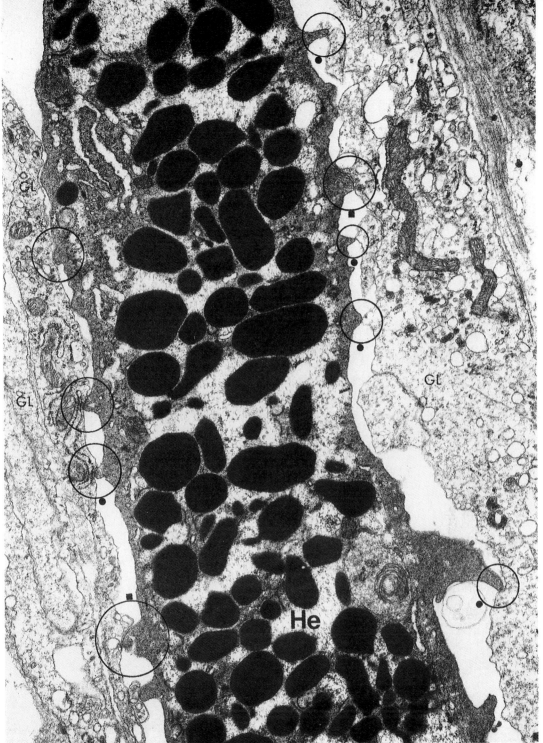

Fig. 159 – Haemocytes (He) from crayfish *Orconectes virilis* form cytoplasmic extensions which abut on glial cells (GL) of damaged nerve sheaths. Contacts of haemocyte processes with adjacent cells (circled areas) are characterized by close apposition of cell processes and matrix. The surface area involved in haemocyte-sheath component contacts is highly variable and ranges from punctate contacts (circle with dot) to broad expanses of intercellular apposition (circle with square). × 18 000. (Dr R.R. Shivers, University of Western Ontario.)

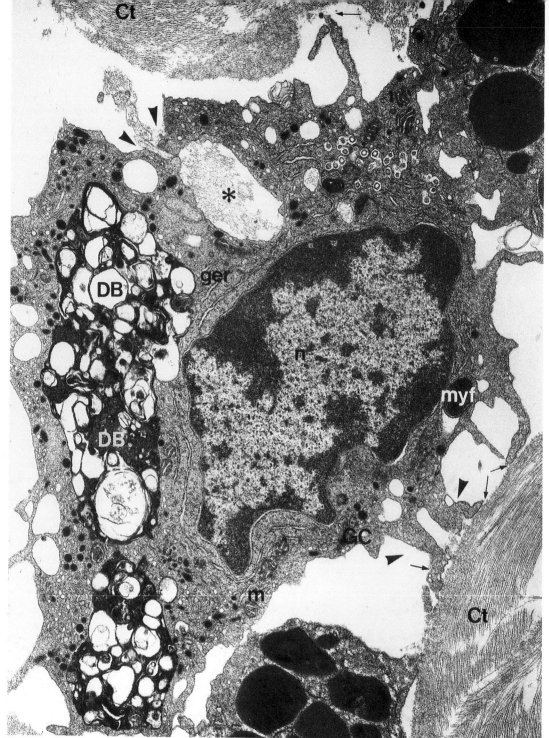

Fig. 160(a) – Haemocyte of the crayfish *O. virilis* associated with components of the glial sheath of damaged nerves. A haemocyte with numerous cytoplasmic extensions is in contact with extracellular matrix (arrows). Filamentous material normally fills extracellular clefts of the intact nerve sheat. Filamentous matrix is present in a forming cytoplasmic vacuole (*) and the cytoplasm also contains two large heterogeneous dense bodies (DB) (lamellar aggregates similar to myelin figures). Several sites on the haemocyte similar to the forming vacuole (*) are suggestive of phagocytosis (arrowheads). × 16 000. (Dr R.R. Shivers.)

myf, myelin figure.

246

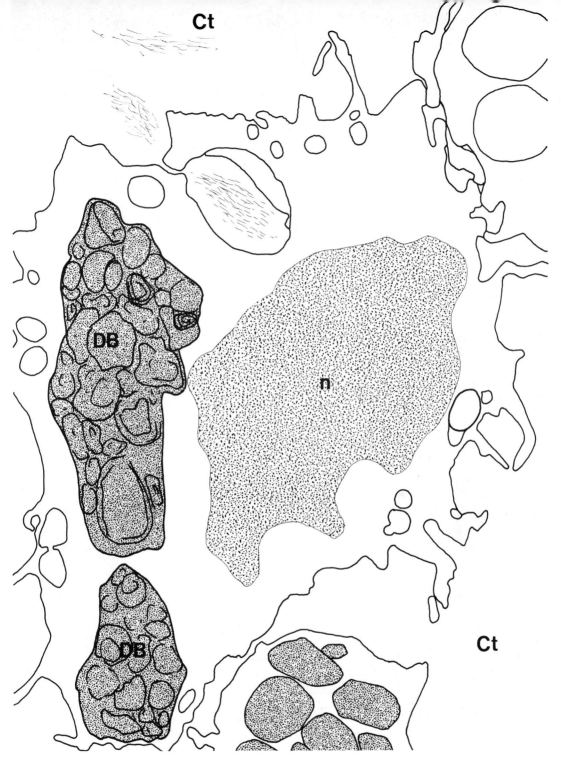

Fig. 160(b) Diagram of Fig. 160(a).

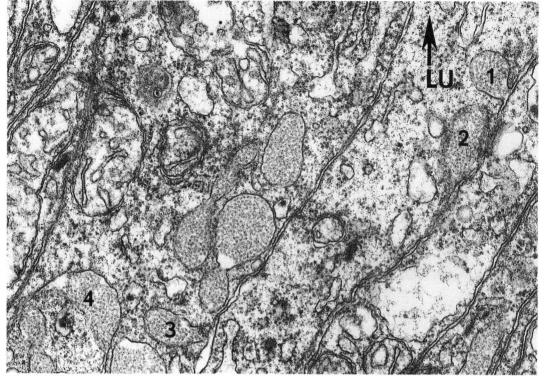

Fig. 161(a) – Mature, haemoglobin-containing secretory vesicles in the apical cytoplasm of heart-body cells of the polychaete *Amphitrite ornata*. Vesicles numbered 1–3 are in the process of secreting haemoblobin into the extracellular spaces. A packet of extracellular haemoglobin molecules is shown at location 4. × 31 000. (Dr M.C. Friedman.)

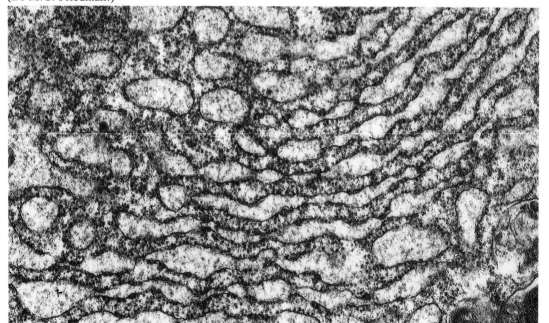

Fig. 161(b) – A portion of the granular endoplasmic reticulum of a heart-body cell of the polychaete *A. ornata*. A moderately dense flocculent material is present within the cisternae of the granular endoplasmic reticulum. × 32 000. (Dr M.C. Friedman.)

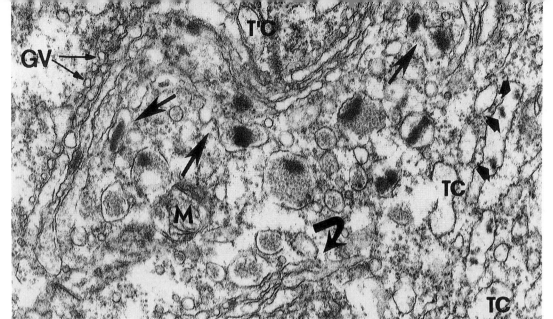

Fig. 162(a) – The Golgi zone of a heart-body cell of the polychaete *A. ornata*. Small, smooth-surfaced transport vesicles (GV), formed at the outer Golgi face may contain material synthesized in the granular endoplasmic reticulum are conveyed to the Golgi complex within partly smooth- and partly rough-surfaced transitional cisternae (TC). The transitional cisternae containing both a moderately dense and a highly dense (arrowheads) material fuse with Golgi membranes (T^{1}C and curved arrow). Condensing vacuoles in which haemoglobin molecules are first identifiable are produced at the inner face of the Golgi complex (straight arrows). × 32 000. (Dr M.C. Friedman.)

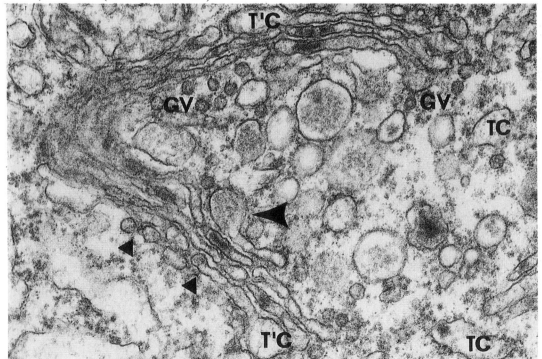

Fig. 162(b) – A portion of another Golgi zone in the heart-body cell of the polychaete *A. ornata*. The transitional cisternae have fused (T^{1}C) to outer Golgi membranes and nearby production of transport vesicles (small triangular arrowheads). A condensing vacuole is apparently budding from the inner Golgi membrane (large arrowhead). × 49 000. (Dr M.C. Friedman.)

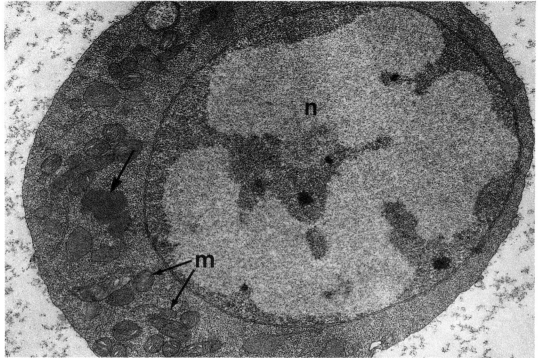

Fig. 163(a) – A very young cyanoblast of the horseshoe crab *Limulus polyphemus*. The cytoplasm contains abundant free ribosomes, small numbers of mitochondria (m) and small crystals of haemocyanin (arrow). Granules in the extracellular space are individual haemocyanin molecules. × 16 500. (Dr W.F. Fahrenbach, Oregon Regional Primate Research Centre.)

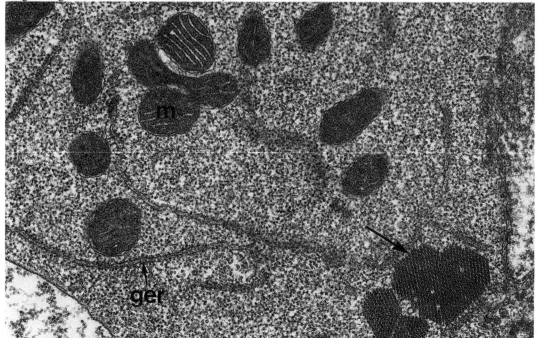

Fig. 163(b) – The cytoplasm of a more advanced cyanoblast of *L. polyphemus*. Granular material is still contained in the lumen of the endoplasmic reticulum but the granules are much smaller than the haemocyanin molecules in the adjacent crystals (arrow). × 46 000. (Dr W.F. Fahrenbach.)

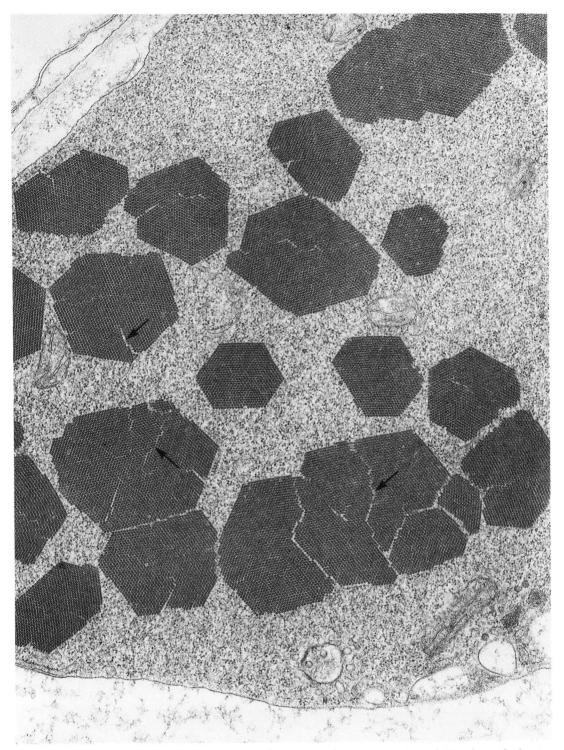

Fig. 164 – A cross-section of haemocyanin crystals in an advanced cyanoblast of *Limulus polyphemus*. Numerous crystalline lattice defects in the form of lacunae and non-conforming planes of fusion (arrows) are present, indicating coalescence of numerous small crystals. × 34 000. (Dr W.F. Fahrenbach.)

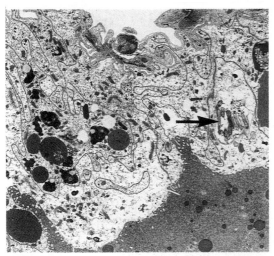

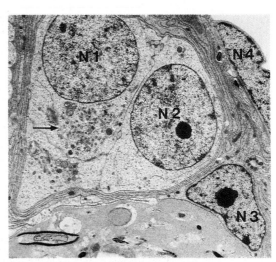

Fig. 165(a) – A portion of dorsally located secretory and phagocytic epithelium within the lumen of the testis of the arachnid, *Cryptocellus boneti*. Large vacuoles contain different inclusions and the flagella of ingested spermatozoa (arrow). × 3 600. (Dr G. Alberti, University of Heidelberg.)

Fig. 165(b) – The germinal part of the testis of *C. boneti*, with germ cells in the early stage of development (stage 1; nuclei N1, N2), cyst with spermatids (stage 5; at bottom), somatic epithelial cells (N3), and the nucleus of a muscle cell (N4). The arrow indicates a concentration of organelles. × 3 600. (Dr G. Alberti.)

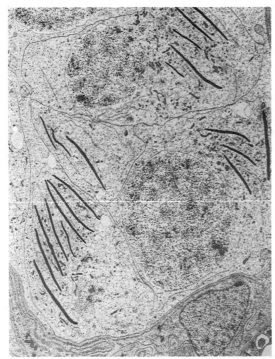

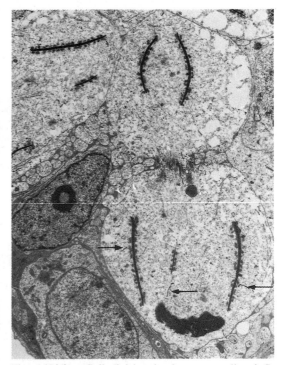

Fig. 165(c) – A cyst of germ cells of *C. boneti* (probably primary spermatocyts) in the stage of meiotic prophase (stage 2). Present in the germ cells are dark plates arranged parallel to each other. At the bottom is the somatic cell region. × 3 600. (Dr G. Alberti.)

Fig. 165(d) – Cell division in the germ cells of *C. boneti*. Mitochondria are attached to the dark plates. Knobbly protrusions occur at the cell surface, especially near cell bridges (electron-dense streaks) (arrows). × 3 600. (Dr G. Alberti.)

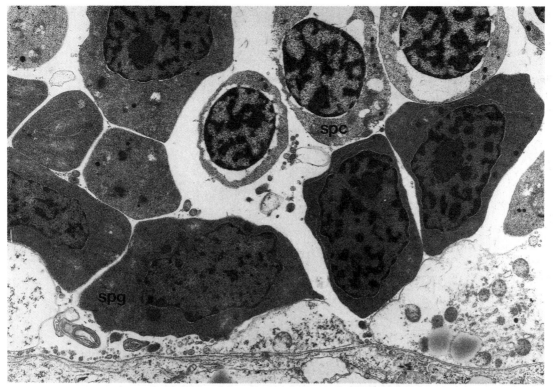

Fig. 166 – Testes of the bivalve, *Mytilus edulis*. Present are spermatogonia (spg) and spermatocytes (spc). × 5 500. (Dr A. Bubel.)

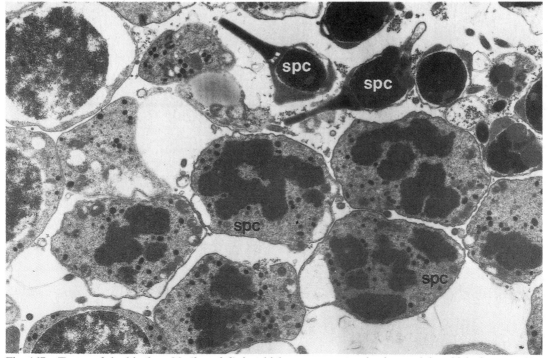

Fig. 167 – Testes of the bivalve, *Mytilus edulis* in which spermatocytes (spc) containing thick chromosomes are in a state of division. × 5 500. (Dr A. Bubel.)

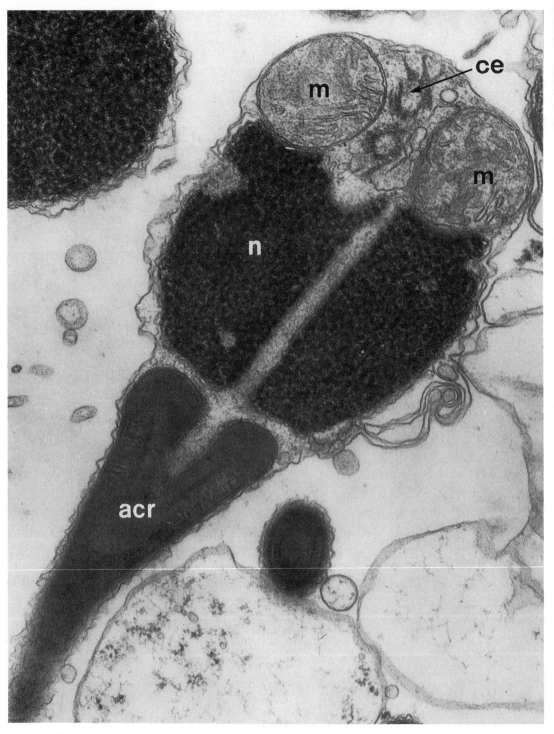

Fig. 168(a) – Testes of the bivalve, *Mytilus edulis*. A late spermatid with a cone-shaped acrosome (acr), mitochondria (m) and centrioles (ce) at the basal portion of the nucleus (n). × 42 000. (Dr A. Bubel.)

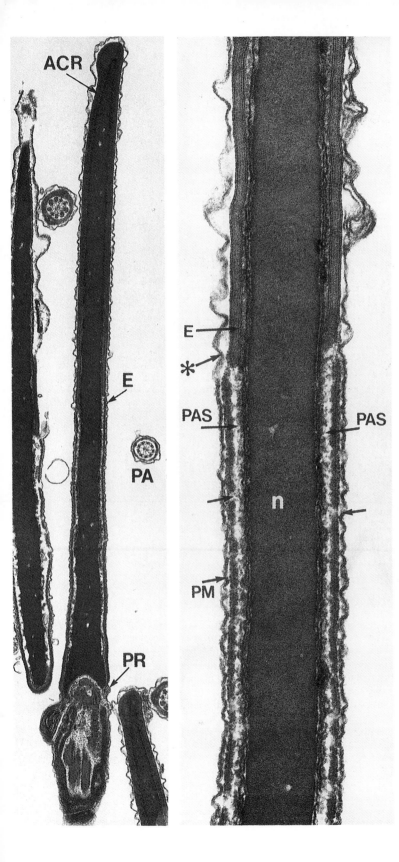

Far left
Fig. 168(b) – A sagittal section of a bovine sperm head. The postacrosomal region (PA) extends from the posterior ring (PR) anteriorly to the distal margin of the equatorial segment (E) of the acrosome (ACR). × 18 000. (Dr G.E. Olson, Vanderbilt University.)

Left
Fig. 168(c) – A sagittal section of a bovine sperm head. The postacrosomal sheath (PAS) is an electron-dense layer immediately under the plasma membrane (PM). In some profiles (*) the sheath appears to contact the equatorial segment of the acrosome (E). There are some discontinuities in the sheath matrix (arrows). Typically the sheath is somewhat separated from the nucleus (N) and the intervening space contains aggregates of electron-dense material. × 22 000. (Dr G.E. Olson.)

255

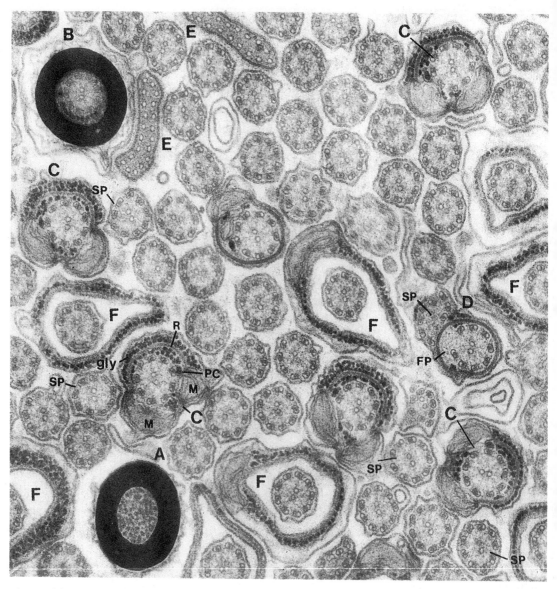

Fig. 169(a) – Transverse section of a group of spermatozoa in the vas deferens of gastropod, *Theodoxus fluviatilus*.

A – apical section of the nucleus; the intranuclear canal is filled with glycogen granules.

B – section through the middle portion of the nucleus; the intranuclear canal is occupied by the axenome.

C – sections of the first portion of the flagellum at its origin. Two mitochondria (M), the spiral ribbon (R) form an electron-dense layer around the 9+2 axial filament.

D – section near the breaking point of the axoneme FP, – first portion of the axoneme; SP, the beginning of the second portion of the flagellum.

E – sections through the flattened tip of the second portion of the flagellum in which there are 20 isolated microtubules.

F – sections through the middle of a typical spermatozoa. × 66 000. (Dr F. Guisti, University of Siena).

PC, protect cylinder

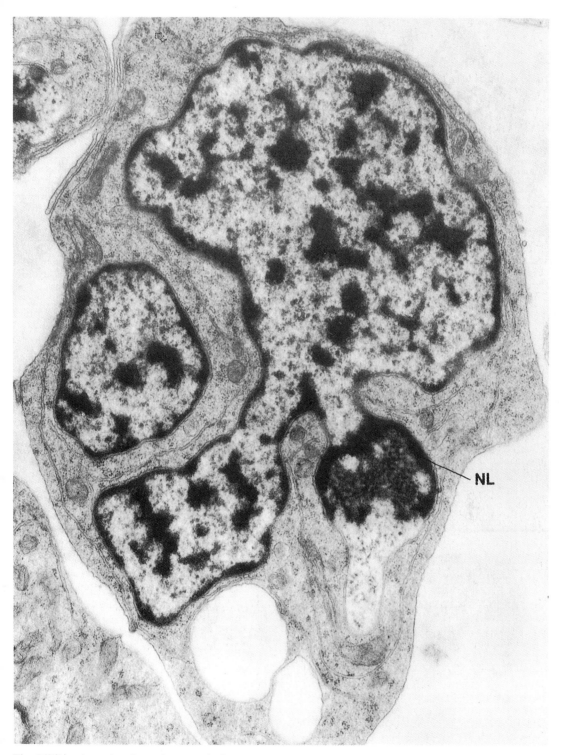

NL

Fig. 169(b) – A section through an atypical spermatozoa of *T. fluviatilius*. The nucleus has its own membrane and possesses irregularly condensed chromatin and lobes (NL). Some of the nuclear lobes separate and move towards the periphery of the cell. × 25 000. (Dr M.G. Selmi, University of Siena.)

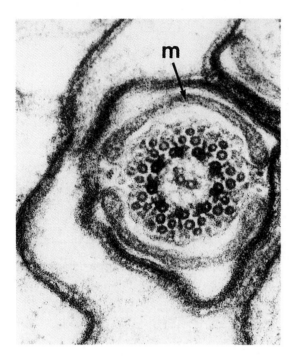

Left
Fig. 170(a) – The mid-piece centred between the crescentic mitochondrial derivatives in the spermatozoon of polychaete *Questa* sp. The 9+2 axoneme is surrounded by nine triads of accessory microtubules. Peripherally directed radial arms separate the triads. × 86 000. (Dr B.G.M. Jamieson, University of Queensland.)

Below
Fig. 170(b) – Spermatozoon of *Questa* sp. The apical member of some of the triads has discontinued, leaving six complete triads. × 94 000. (Dr B.G.M. Jamieson.)
m, mitochondrion.

Right above
Fig. 170(c) – Cross-section of a young spermatid of the insect, *Neocondeelum dolicharsum*, revealing the position of the 13+0 axoneme (A) in a nuclear groove. × 34 000. (Dr R. Dallai, University of Siena.)

Right below
Fig. 170(d) – Cross-section of a posterior end of a young spermatid of *N. dolicharsum*. The 13+0 axoneme (A) formed by microtubular doublets devoid of arms, is flanked by a mitochondrion (m). × 130 000. (Dr R. Dallai.)

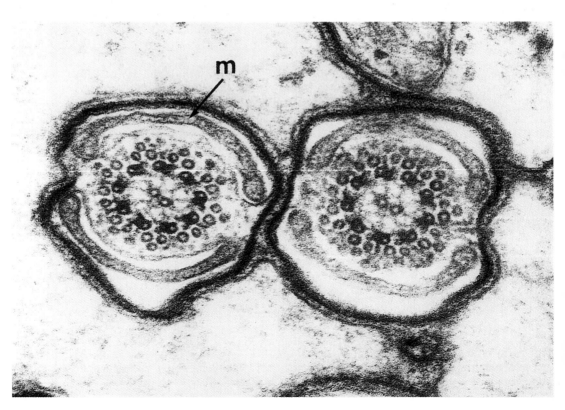

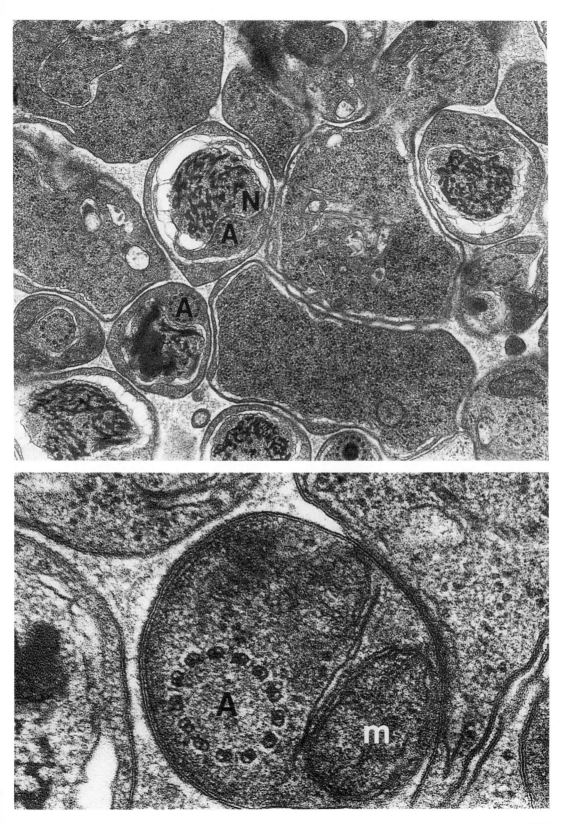

259

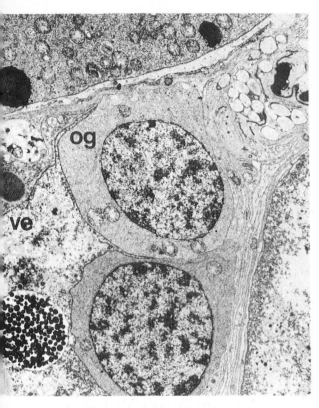

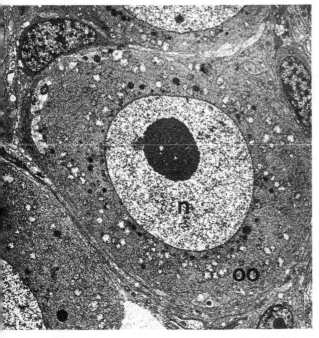

microtubules typically remain. However, the axoneme of insect, *Neocondeelum dolickarusum* is of the 13+0 type (Fig 170c, d).

Oogenesis

In females, the ovum or egg is produced within follicles of the ovaries by a process called oogenesis. In the follicle, the parent cell of the future ovum is called the oogonium. The oogonia give rise to primary oocytes, which enlarge and undergo unequal meiotic division to form a large secondary oocyte and a very small polar body. The nucleus of the secondary oocyte enters the second meiotic division and produces a large cell that differentiates into the ovum and a second polar body. The oogonium possesses a large nucleus with a prominent nucleolus and a cytoplasm characterized by numerous free ribosomes (Fig 171a). The growing follicle consists of several cellular and acellular layers. The oocyte within the follicle is immediately invested by an acellular envelope which contains microvilli processes that extend from the oocyte (Figs 172a, 173). At later stages in development, oocytes show a striking increase in nuclear and nucleolar volume, and a noticeable differentiation in the cytoplasm (Fig 171b). The numbers of organelles, such as granular ER, mitochondria, Golgi complexes, lipid and glycogen increase in number in the greater cytoplasm volume. The nuclear envelope blebs at this stage and this may be an indication of nucleocytoplasmic exchanges. (Fig 174). In coelenterates, cytospines start to appear on the oocyte surface. Cytospines resemble large microvilli (Fig 23a). In many invertebrates vitellogenesis may be the dominant activity of the oocyte for much of its period of existence in the gonad. Vitellogenesis is the accumulation within the oocyte of reserve material for use after fertilization. During this stage, the ER increases, annulate lamellae appear and yolk granules are formed. Within larger oocytes, the vitelline membrane increases in thickness and the oocyte continues to form yolk granules. Most characteristically, the oocyte appears, in some species, to sequester material by pinocytosis (Fig 172b). Proteins from blood capillaries may traverse the extracellular spaces of the follicular epithelium and vitelline membrane and enter the oocyte by micropinocytosis, to be used in the formation of yolk granules.

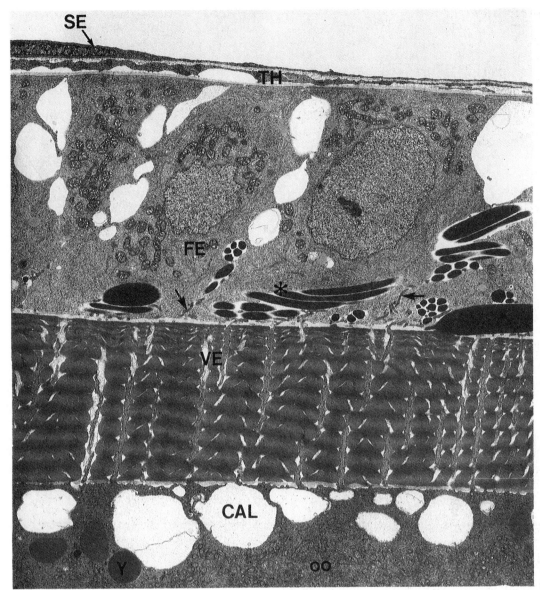

Left above

Fig. 171(a) – Two adjacent oogonia (og) of the platy-helminth *Vorticeros luteum* with poorly differentiated cytoplasm. In the upper area, part of a vitelline envelope (ve) with a cocoon shell globule is present. A small portion of a presumed mature oocyte is evident on the right. × 5 000. (Dr N. Nigro, University of Pisa.)

Left below

Fig. 171(b) – A differentiating oocyte (oo) of *V. luteum*, surrounded by a cytoplasmic accessory cell projection, containing small electron inclusions in the perinuclear cytoplasm. × 2 900. (Dr N. Nigro.)

Above

Fig. 172(a) – The outer region of the follicle of the sheeps-head minnow, *Cyprinodon variegatus*. The oocyte (oo) is surrounded by an elaborate vitelline envelope (VE), which has perpendicular channels containing extensive microvilli, which project from the oocyte surface. The follicle cells are separated from one another by intercellular channels, which contain the long microvillar processes of the oocyte (arrows) and the fibrillar appendages of the vitelline envelope (*). A relatively thin thecal layer (TH) lies between the follicular epithelium (FE) and the attenuated layer of surface epithelial cells (SE). A cortical alveolus (CAL) and a small yolk sphere (Y) are evident. × 6 000. (Dr K. Selman, University of Florida College of Medicine.)

261

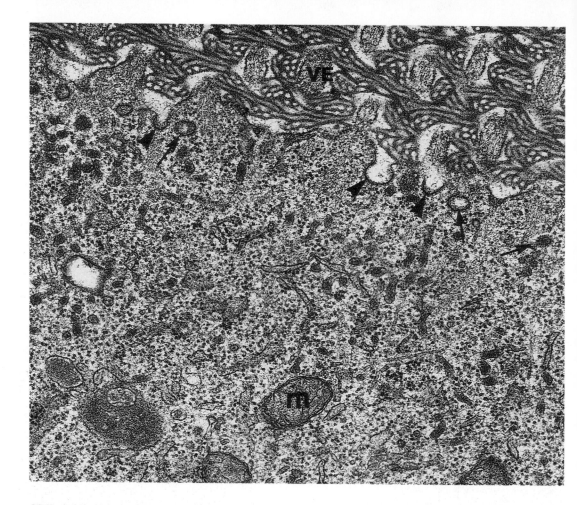

CHLOROPLASTS

Chloroplasts are the most obvious of the cytoplasmic organelles present in the cells of plants. They are relatively large, being about 5–10 μm in diameter, and from 2 to 3 μm in width. A mature chloroplast is bounded by an envelope composed of two unit membranes. However, unlike mitochondria, the internal membrane system does not appear to be connected with the envelope. The membranes with the envelope form lamellae or thylakoids, which consist of two membranes, each about 7 nm thick, lying adjacent to each other, but separated by an intramembrane space which may vary from 4 to 7 nm in width. These thylakoids run throughout the matrix or stroma of the chloroplast. In places these lamellae are stacked to form grana, which are like piles of coins (Fig 1b, c, 175). The grana are about 0.3–2 μm in diameter and in general there are

Above

Fig. 172(b) – The surface of a *C. variegatus* oocyte from an early vitellogenic follicle in which abundant endocytotic activity is apparent. Coated pits (arrowheads), coated vesicles (arrows) and smooth-surfaced tubules are numerous at the oocyte surface. × 36 000. (Dr K. Selman.)
VE, vitelline envelope.

Right

Fig. 173 – Ovary of bivalve, *Mytilus edulis*. A section through the oocyte depicting the vitelline envelope (VE), microvilli (mv), lipid droplets (l) and yolk granules (y). × 9 000. (Dr A. Bubel.)

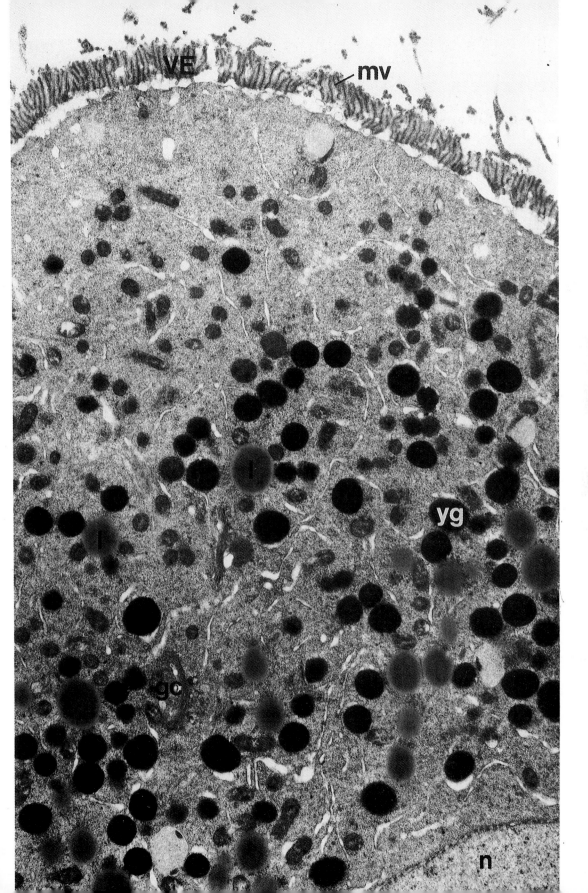

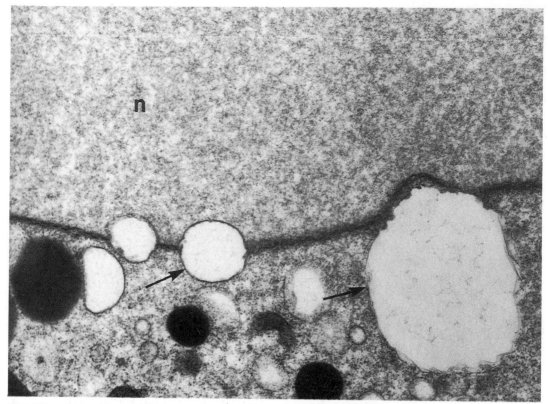

Fig. 174 – Ovary of the bivalve, *Mytilus edulis*. There is nuclear blebbing (arrows) in the nucleus of the oocyte, apparently indicating nucleocytoplasmic exchanges. × 30 000. (Dr A. Bubel.)

between 10 and 100 thylakoids per granum and 40–60 grana per chloroplast in typical photosynthetic cells. Chlorophyll is located on both stromal and granal thylakoid membranes. The green colour of many leaves and stems in plants is due to the green pigment, chlorophyll, of chloroplasts. The close association of the chloroplast and its green pigment is one of the most significant in nature. The energy from sunlight is first captured by chlorophyll and eventually becomes trapped in chemical bonds of organic molecules, primarily sugar.

Environmental factors have a marked influence on chloroplast and thylakoid structure, and also on thylakoid composition. Light conditions during plant growth affects both thylakoid composition and chloroplast structure. However, when light is not limiting, the stage of development of the plant seems to have a significant effect on the thylakoid components and structure. During ageing, chloroplasts are found to show an increase in the size of the grana and distinct profiles of stroma thylakoids extending between large grana.

Small double-membrane-bound bodies that appear to be modified bacteria exist in chloroplasts of freshwater dinoflagellates, as endosymbionts. The chloroplast symbionts are between 1 and 1.5 μm in diameter and contain thin filaments of 'naked' DNA and putative prokaryotic-sized ribosomes. The outer of the two membranes that surround the symbiont may be expanded to form cisternae, which frequently connect with the outer membrane of adjacent symbionts (Fig 176a, b).

Chloroplasts, like mitochondria, have their own DNA and ribosomes and are self-replicating. A plant cell will thus pass on to its offspring only the types of chloroplasts that it carries in its own cytoplasm. However, unlike mitochondria, which are involved primarily with catabolic or breakdown processes, chloroplasts are concerned with anabolic or building-up processes.

264

SIEVE TUBES

In plants, sieve tubes are conducting cells and consist of rows of elongated cells with perforated end walls known as sieve plates. These conducting cells are essential components of the phloem which provides the major pathway for the transport of sugars and other photosynthates from the leaves to other parts of the plant. When mature, sieve tubes contain no nuclei and are associated with smaller cells, called companion cells, which contain nuclei. The earliest detectable sieve elements resemble parenchyma cells in having cytoplasm rich in scattered ribosomes, clearly defined tonoplast, a small amount of ER, conspicuous Golgi complexes, some mitochondria and relatively undifferentiated plastids (Fig 177a). The cell wall separating the protoplasts of two adjoining future sieve elements is traversed by plasmodesmata (Fig 177a). Plasmodesmata are an interesting feature of plant cells. They are threads of cytoplasm which are lined by plasma membrane and extend through the cell wall connecting adjacent cells. The pores are usually 50–80 nm in diameter. The function of plasmodesmata is not understood, although they may provide a major pathway for the movement of materials and for communication between cells. In plasmodesmata, both the plasma membrane and the ER are continuous through the plasmodesmal canal. A pore develops by enlargement of a plasmodesmal canal and removal of a plasmodesmal core. Initially, callose is deposited at each end of a plasmodesma, encircling its strand. Callose accumulates and thickens the wall at the pore site, which is covered with a plasma membrane and ER cisternae.

Lysing of the wall middle-lamella region allows the merging of the two opposite callose masses. The removal of the plasmodesmal strand and most of the callose opens the pore, which is lined by the plasma membrane (Fig 177b).

BACTERIA

Bacteria are prokaryotic cells surrounded by a rigid cell wall, 10 nm or more thick, containing protein, polysaccharide and lipid. Inside the cell wall is a plasma membrane. However, in the cytoplasm, most of the cytoplasmic organelles present in eukaryotic cells are lacking. Bacteria have no nuclear envelopes, a nuclear region containing a single circular molecule of DNA is present instead. Surrounding the DNA are ribosomes, which are found in groups called polysomes or polyribosomes. In bacteria there are no membraneous structures such as ER, Golgi complexes, lysosomes and mitochondria. The enzymes involved in oxidation that constitutes the respiratory chain are

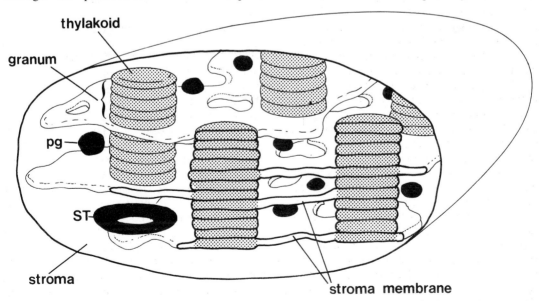

Fig. 175 – Diagram of plant chloroplast.   pg, pigment granule; st, starch granule.

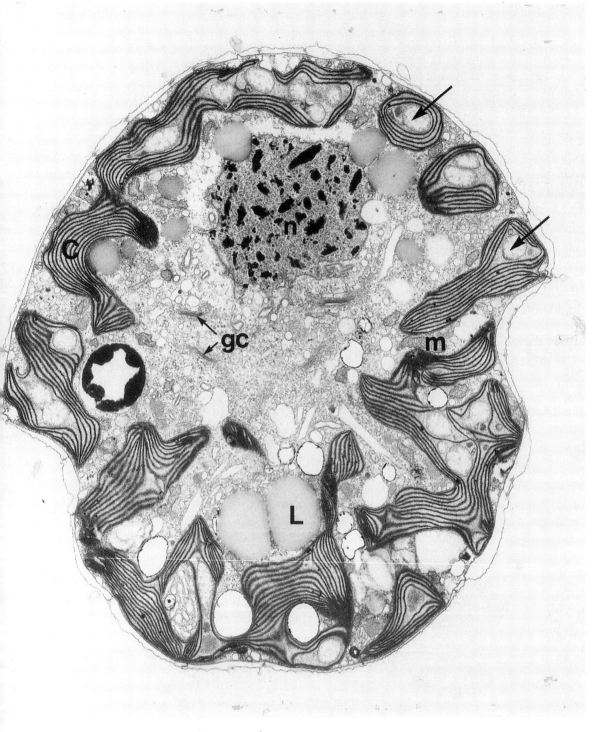

Fig. 176(a) – Longitudinal section through the dinoflagellate, *Woloszynska pascheri,* revealing the nucleus (n), lipid droplets (L), Golgi complexes (gc), mitochondria (m) and a reticulate peripheral chloroplast containing profiles of approximately 20 endosymbionts, two of which are indicated by arrows. × 7 300. (Dr L.W. Wilcox, Ohio State University.)

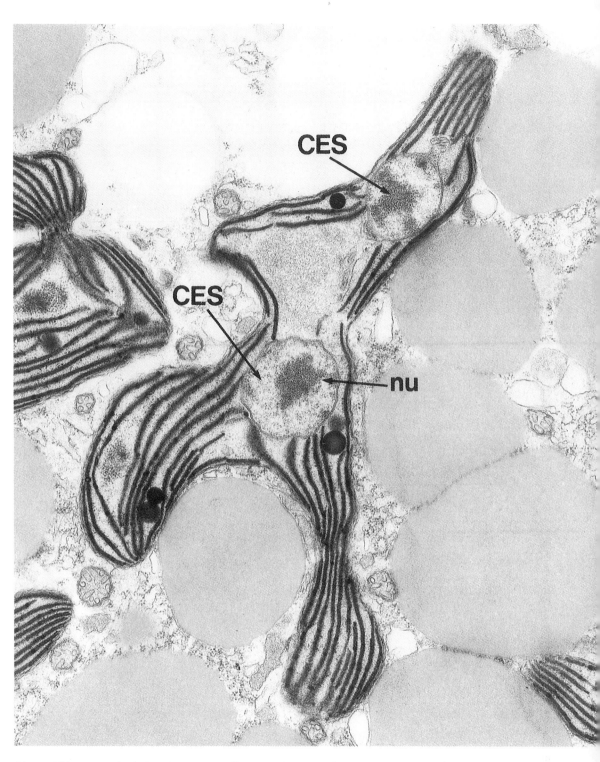

Fig. 176(b) – A section of a planozygote of *W. pascheri* containing two chloroplast endosymbionts (CES) whose ribosomes are found in dense clusters resembling nucleoli (nu). × 25 000. (Dr L.W. Wilcox.)

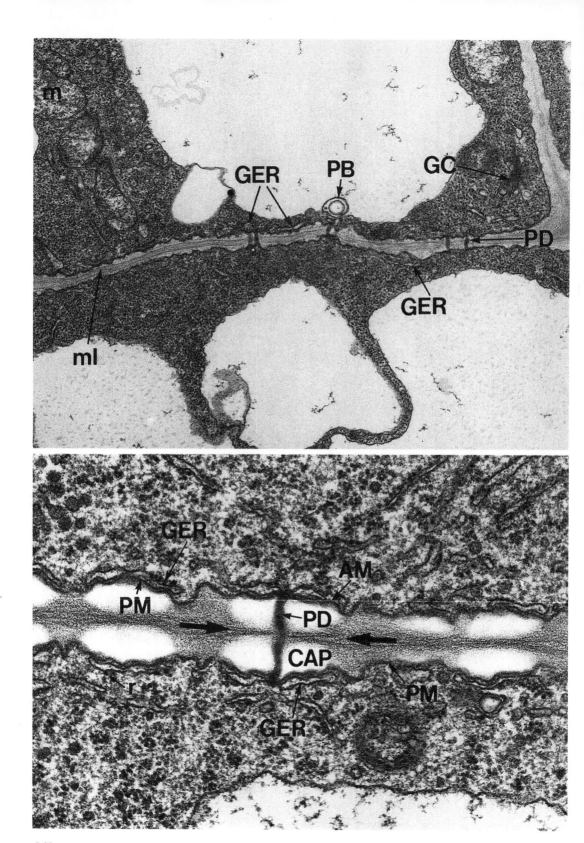

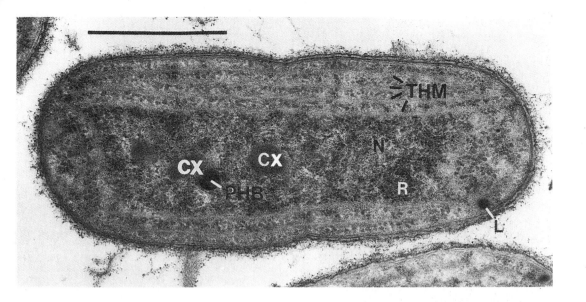

Left above

Fig. 177(a) – Longitudinal section of parts of two sieve elements of *Echium angustifolium* separated by a differentiating sieve plate in which the cell wall is penetrated by six plasmodesmata (PD). Two of the latter are especially close to each other. Both sieve elements have intact tonoplasts. × 24 000. (Prof. K. Esau, University of California.)

ml, middle lamella; pB, paramural body.

Left below

Fig. 177(b) – Longitudinal section of a differentiating sieve plate of *E. angustifolium*. One pore is sectioned through the plasmodesma (PD) which extends through two callose platelets (CAP) and the remainder of middle lamella (large arrow). Plasmalemma (PM) lines the wall on both sides covering the callose platelets. Endoplasmic reticulum (GER) cisternae appear as two profiles on each of the pore sites, including the plasmodesma. The space between endoplasmic reticulum and plasmalemma is filled with alveolar material (AM). × 60 000. (Prof. K. Esau.)

r, ribosome.

Above

Fig. 178(a) – Longitudinal section through the cyanobacterium, *Agmenellum quadruplicatum*. The peripheral thylakoid region indicates a photosynthetic thylakoid membrane (THM) and lipid bodies (L). The thylakoid system surrounds the central cytoplasmic zone, which includes carboxysomes (CX), nuclear material (n), polyphosphate bodies (PHB) and ribosomes (R). × 39 000. (Dr S. E. Stevens, Pennsylvania State University.)

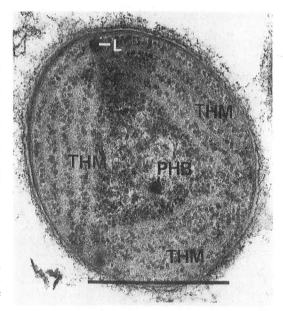

Fig. 178(b) – Cross-section through the cyanobacterium, *A quadruplicatum*. Thylakoid membranes (THM) form a pattern of concentric triangles that tend to intersect at the corners. × 39 000. (Dr S. E. Stevens.)

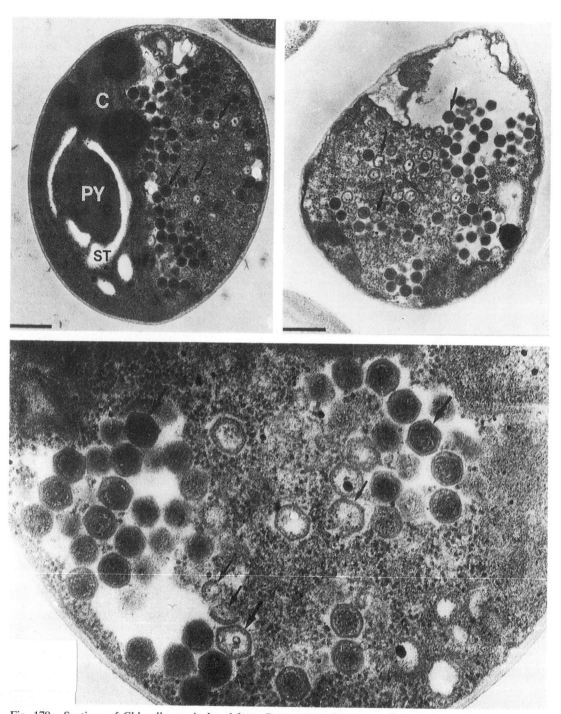

Fig. 179 – Sections of *Chlorella* sp., isolated from *Paramecium bursaria* and grown in sterile mass culture, after infection with a specific *Chlorella*-virus (CVM-1). Arrows indicate different phases of development. (a) Intermediate phase of infection. The chloroplast is intact with an assembly of capsids with less heavily stained contents around a virus-free part of the algal cell (nuclear region?) × 26 000 C, chloroplast; PY, pyrenoid; ST, starch. (b) Late phase of infection. The chloroplast is disrupted. × 26 000. (c) Intermediate phase of infection. Small arrows indicate incomplete capsids (precursors?) × 79 000. Dr. W. Reisser, University of Lahnberge.)

associated with the plasma membrane. There is an exception to the rule that prokaryotic cells lack intracellular membraneous structures. A photosynthetic bacterium contains chlorphyll, which is associated with membraneous vesicles or lamellae. The lamellae are not contained in membrane-bound plastids (Fig 178a, b). Some bacteria have hair-like processes, about 10 nm in diameter and of variable length, called flagellae. In contrast to the flagella and cilia of eukaryotic cells, which contain several fibrils, each flagellum in bacteria is made up of a single fibril.

VIRUSES

Viruses are associations of two macromolecular components, an outer 'coat', or capsid, of protein, and an inner 'core' of nucleic acid, either DNA or RNA, but not both. Viruses function by invading cells. Inside the host's cell, the virus reproduces itself by using the cell's chemical machinery. Often, this destroys the host cell. The protein coat or capsid may be complicated, with a tail and long leg-like fibres, or it may be a simply polyhedron or rod. Many animal viruses, some plant viruses and a few bacteriophages have a membranous envelope surrounding the capsid. This envelope is derived from the plasma membrane of the host cell in which the virus was produced. A tail-less polygonal virus with a prominent capsid of about 140–150 nm in diameter and about 14–15 nm thick, is found in pond water. This virus shows a marked host-specificity in attacking only endosymbiotic *Chlorella sp.* isolated from *Paramecium bursaria* (ciliate) (Figs 179a–c).

CELL DEATH

Cells undergo degenerative changes which lead to cell death. Such cells exhibit disrupted membranes, a decrease in cytoplasmic density, swollen mitochondria, a distended and fragmented ER and nuclei with condensed chromatin and dilated nuclear envelopes (Fig 180).

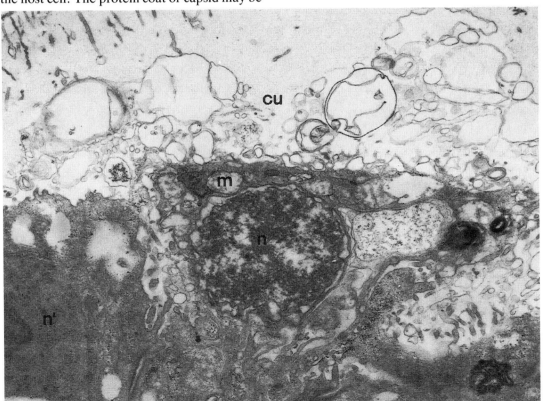

Fig. 181 – Cell death in the opercular epithelium of the polychaete *Spirorbis spirorbis* adjacent to a wound. There are abnormalities in the nucleus, mitochondria and cuticle (cu). × 12 000. (Dr C.A.L. Fitzsimons.) n[1], normal nucleus.

INDEX
Page numbers in *italic* refer to figures.